ATOMIC SUICIDE?

Dr. Walter Russell

Lao Russell

ATOMIC SUICIDE?

BY

DR. WALTER AND LAO RUSSELL

Illustrations by

Dr. Walter Russell

Introduction by

Lao Russell

UNIVERSITY OF SCIENCE AND PHILOSOPHY

FORMERLY THE WALTER RUSSELL FOUNDATION

SWANNANOA, WAYNESBORO, VIRGINIA

First Edition 1957 10,000

PRINTED IN THE UNITED STATES OF AMERICA

For the preservation of humanity in its threatened hour of total extinction this book is humbly dedicated.

THE SECOND EPISTLE OF PETER

Chapter 3–12

"Looking for and hasting unto the coming of the day of God, wherein the heavens being on fire shall be dissolved, and the elements shall melt with fervent heat."

The whole issue as to whether radioactivity can or can not be used as the next world fuel, stands or falls upon one issue alone, and that issue is whether science can protect organic life from extinction, or whether it can not. Our contention is that it can not.

THE AUTHORS

Contents

Introduction

An Open Letter to a Fearing World Concerning our Authority for the New Scientific and Cosmic Knowledge Given Herein.

By

LAO RUSSELL

DEAR READER:

A very heavy responsibility rests upon the shoulders and conscience of anyone who would dare write such a book as this at a time when the world is in dire need of what nuclear fission promises. We fully realize that fact, but over and above it lies the greater fact that loving service to one's fellow man comes first, before all things else.

This universe is founded upon love. Man's actions must, therefore, be founded upon the principle of love. We have no motive for writing this book other than the motive of service to our fellow man. We have new knowledge which is purposeful in the building of an enduring civilization upon the present one that has been rapidly declining since the year 1900 toward another long Dark Age period. During that fifty year period it has fallen farther than it had formerly risen in eight centuries.

The wars of the last forty years have been the main cause for developing nuclear fission. Our fuel need was not primary. It was secondary, yet the development of it continues with seeming full assurance that its admitted dangers can be voided. We know that they cannot be voided, and can tell our fellow man the cosmic reason *why* their use, as now intended, would make of this earth a barren, oceanless rock for many thousands of years.

That new knowledge was not given to us to keep. It was given to us to regive in service to our fellow man. It is in that light of recognition of the universal brotherhood of man that we now give it, with the hope that it will not only deter man from his intended self-annihilation, but will also reverse the entire practice of human relations which is now heading mankind for another fall; and save him by thus giving him a better way to fulfill his One World ideal.

We also believe that the new knowledge concerning our universe, and man's closer relation to God than he has ever suspected, will not take away the needed fuel he now dangerously seeks but will give him the better, and more abundant safe fuel, which Nature intended for him at this stage of his unfolding.

Our knowledge is divinely inspired and purposeful for this day and age. All epoch making knowledge is divinely inspired. You could not possibly read this book without being divinely reinspired into recognizing its fundamental truths, even though they may be a thousand years beyond this day's thinking, and, consequently, very contrary to it.

In love for our fellow man, and our beautiful earth, we present this solution of long hidden mysteries of the ages for your deep *inner thinking* in quiet meditation.

I

Many world leaders sincerely believe that nuclear fission can be safely used for industry, and that mankind can be protected from its admitted dangers. We believe, however, that such use would destroy all organic life on this planet for many centuries.

When you acquired this book a heavy question undoubtedly raised itself in your consciousness as to our authority for daring to write a book, which questions the wisdom of atomic energy for world use, either for war or industry. The acquisition of it by you also indicates that you are a *thinker* who will read the contents with an open mind.

Within these pages lies the dynamic answer to "the riddle of the universe," which will give to man the knowledge and power to control all matter and motion, and to solve the secret of life and death, and growth and decay, for which man has sought but never found. *As long as man depends upon the evidence of his senses for knowledge of CAUSE of the EFFECTS—which are the limit of his vision—he will never solve these secrets.* Up to this day in human history empirical knowledge, which means knowledge gained by sensed-observation, has been deemed the only reliable evidence upon which to form conclusions. This assumption is a colossal error of judgment, for not one effect in Nature is what it *seems* to be. Man will never find the answer to the secrets of the universe that way.

The time has come, however, when man must have the right answer, else all men will perish from this planet. This tremendous statement is not made lightly. My husband and I well know the ridicule which would result from our inability to support such a seemingly fantastic claim to transcendent knowledge. We would not make such a claim, however, nor would we write this book now, except for man's dire need in this most crucial day of his history—*and every moment lost could be man's last moment on earth.*

As briefly as possible, I will give you our authority which will be proved step by step in every page of this book. We fully realize that our authority is in ourselves. You must, therefore, first know about us. That would be your first question, so I will first answer this for you.

My husband, Dr. Walter Russell—who at this time is 86 years of age—is no ordinary man. Before he was thirty he had gained an international reputation, with high honors from nine European countries for masterliness in art. Judging him by the world's standards he could be called the man who has achieved the most in all the world in his various forms of endeavor, which has won for him the title of being the most versatile man in America.

A quick look at his background will justify this seemingly extravagant claim. Without having had any schooling above the primary grade in a country school, he has become a master painter, sculptor, architect, composer and musician, author, philosopher, humanist and scientist, having excelled in every one of these fields.

At thirteen he was a church organist. Before the age of sixteen he had composed over a hundred melodies, one of which so pleased Paderewski that he had it put into manuscript form and invited my husband to play it with the Boston Symphony Orchestra at one of its rehearsals. At nineteen his illustrations excited the interest of Mark Twain from which a lifelong friendship grew and The Mark Twain Memorial became my husband's greatest masterpiece in sculpture. Other famous memorials by him are The Four Freedoms, The Colin Kelly Memorial and the Charles Goodyear Memorial. At the age of thirty he was acclaimed a master portrait painter of children, and was commissioned to select and paint the twelve most beautiful children in America by The Ladies Home Journal.

Long before that, in 1895, my husband became leader in the great humanist movement started in 1875 by Herbert Spencer, Ralph Waldo Emerson, Walt Whitman, Edwin Markham, John Burroughs, Oliver Wendel Holmes, Andrew Carnegie and Richard Watson Gilder, which was first known as The Twilight Club, until Alexis Carrel called it The Science of Man, which it now is, and of which we are still its leaders and teachers. It was incorporated as The Society of Arts and Sciences during the twelve years of Thomas Watson's and my husband's association in spreading its ethics and morals throughout the business world.

Before Mr. Watson played a leading role in this great moral movement, my husband, in association with Theodore Roosevelt, Dan Beard, Rudyard Kipling, Sir Arthur Conan Doyle and Edwin Markham instituted the Boy Scouts movement in England through Lord Baden Powell, and in America, through Dan Beard.

The Spencer moral code was based upon the righteousness of good citizenship and service to man. Great fruits soon grew out of my husband's leadership, such as Rotary Club, Kiwanis, Lions and other service clubs. Sweat Shop Elimination, Compulsory Schooling, Advertising Censorship and Better Business Bureau soon followed with the help of Adolph Ochs, owner of The

New York Times, who worked with my husband for years to infiltrate this new note of humanism and morality into industry. A list of the great educators and scientists, men who were associated with my husband during the more than forty years of his active work for the betterment of human relations in industry, would fill many pages. However, the list includes such men as Dr. Nicholas Murray Butler, Hamlin Garland, Calvin Coolidge, Richard Le Gallienne, Mark Twain, John Dewey, Henry Holt, Robert Collier, Howard Chandler Christy, Edwin Markham, James Howard Bridge, Rudyard Kipling, George Gould, Louis Tiffany, Ernest Thompson Seton, Nikola Tesla, Thomas Edison, Walter Damrosch, Michael Pupin, Theodore Roosevelt, Dr. Lee de Forest, Adolph Ochs, A. Cressy Morrison, Robert Andrews Millikan, Thomas J. Watson, Dr. Henry Norris Russell, Otto Kahn, J. Pierpont Morgan, Dr. Albert A. Michaelson, Cass Gilbert, Gen. Coleman Du Pont, Douglas L. Elliman, James A Mears, Whitney Warren, George Gray Barnard, Thomas Hastings, Mrs. Harry Payne Whitney, Cornelius Vanderbilt, Oswald Spengler, Alexis Carrel, Eugene Grace, Irvin S. Cobb, Irving Bacheller, John Phillip Sousa, Victor Herbert, Charles De Kay, Sophie Irene Loeb, August Hecksher, Dr. H. H. Sheldon, Dr. Harvey Rentschler, Gerard V. Swope and many others. Letters, autographs and other historical documents from many of these celebrities are part of the museum of my husband's works here at Swannanoa.

A record of the moral uplift in the business world, during the twelve years in which Mr. Thomas Watson and my husband were so closely associated in this work, would also be strongly felt in the standards of business and social morality, if it could be visibly recorded with the strength in which it has been demonstrated.

My husband's love of beauty prompted him to demonstrate it by infiltrating the idea of beautiful Parkways, to by-pass towns and cities, among civic leaders. The Bronx River Parkway became the first fruit of this inspiration, which has spread around the world. As an award to the engineers of the Bronx River Parkway, who became inspired by this idea, my husband bestowed gold and bronze medals upon them during his seven years presidency of The Society of Arts and Sciences. Copies of these medals have also become a part of the permanent collection at Swannanoa.

To enumerate all of his achievements would require a volume rather than a few pages. Very briefly, therefore, I will tell you that he has designed about eighty millions of dollars in value of architectural structures, civic centers and apartment buildings, without ever having studied architecture or engineering. In New York alone, he erected thirty million dollars worth of studio and apartment buildings, including the whole artist colony on West 67th Street where he lived in one of them, Hotel des Artiste's, for many years. He built the first Hotel Pierre, at 290 Park Avenue, and created the cooperative movement in New York. He also created the duplex, triplex and penthouse apartment idea.

He became official painter for Theodore Roosevelt, Governor Ames of Massachusetts and many other well known families. He had never touched sculpture until the age of 56, when he became official sculptor for Thomas Edison, Mark Twain, Franklin D. Roosevelt, Victor Herbert, John Phillip Sousa and many others. As an author his many books include The Bending of the Twig, The Age of Innocence, The Sea Children, The Universal One, The Russell Genero-Radiative Concept, The Secret of Light, The Book of Early Whisperings, The Divine Iliad and The Russell Cosmogony.

In science Dr. Russell was the first to give knowledge of the existence of plutonium, neptunium, deuterium and tritium to the world—which we shall refer to later—although these were not the names that he originally gave to these hitherto unknown elements, as you will observe in the accompanying charts.

It is no wonder, therefore, that he has so often been called The Modern Leonardo, and Dean of the Arts. It is, also, a strange human characteristic to belittle, suppress and even crucify such men who are so far ahead of their time. Such great souls are the world-mystics whom the man who does not understand calls "crackpots" and dreamers. Practically every world-genius has been a "crackpot" to his neighbors. Poor Leonardo had to bury his inventions in sealed vaults, as Nikola Tesla told my husband he must also do, so that posterity could take advantage of his "crackpot" ideas and inspirations.

My husband has demonstrated the same standard of versatility which Leonardo's whole life demonstrated. Those thousands who have visited Swannanoa—our mountain-top home in Virginia—have ample proof of this, for here I gathered together all that I could of his works in science, which he was preparing for posterity, and in all of the arts—of which he is a complete master—to prove to the world that all-knowledge and all-power lies within every man who becomes aware of that divine power within him.

An outstanding characteristic of his is the fact that he was never tired during his whole life—until very late in it. No matter how hard he worked, physically or mentally, he had within him an inner joyousness, which made fatigue impossible until sleep claimed him.

My husband is a living example of the complete supremacy which every man can acquire who works and lives KNOWINGLY WITH GOD with complete moment to moment awareness of His presence. With such a power extended to him from the universal Source he became master of every art he desired to express himself in. He even demonstrated this in his sports, by earning the National Championship for figure skating and in becoming an acknowledged master in the training of Arabian stallions for high school exhibition, and for fine saddle horses.

The little book written by the late Glenn Clark about my husband which was named: "THE MAN WHO TAPPED THE SECRETS OF THE UNIVERSE," is the most apt title possible, for there has never been a man who could more ably live up to that title.

Many of you have seen the visible proof of his works of art and know countless other things he has accomplished, but the world as a whole is not aware of his greatest of all contributions to the world of science. Let me briefly recount some of these for you so that you may become aware of his great contributions to science, and know the Source from which he gained this new vital scientific knowledge that we have written in "ATOMIC SUICIDE?"

Dr. Russell tried to give his scientific knowledge to the world in 1926. He issued two charts which contained three invisible space octaves preceding hydrogen, and also the transuranium elements which are beyond uranium, which are now the basis of radioactive fission. He sent these charts to all the world of science without receiving one letter of recognition for this greatest of all contributions to science since Medeleef gave the basis for it in his incomplete table of the elements. We herewith print these two charts which have made so profound a change in scientific practice, *including the atomic age which resulted thirteen years later,* and the discovery of the incorrectly named "isotopes" of hydrogen. *These are not isotopes, however, for my husband had placed them there as full toned elements—as they are.* Take note that the new names given by the laboratory workers who isolated them are placed beside their positions and their copyrighted names in the charts. See Figures 1 and 2.

Not until 1941, after heavy water, deuterium, tritium, neptunium and plutonium were isolated in some of the many laboratories, which my husband's charts—plus three years of lecutres—caused to come into being, did he receive any recognition for this contribution. He was then awarded a Doctor's Degree of Science for these very charts by The American Academy of Sciences, an organization which gave degrees to people for work performed, such as an Edison or Marconi, who could not acquire one from a university because of not having studied science in a university.

These priceless new elements, which justly give him the right to be recognized as the father of the atomic age, were not given the names which he had copyrighted when they were finally isolated. For this reason my husband has never been accredited as the man who first gave knowledge of their existence to the world.

He labored hard for years in urging such research laboratories as Westinghouse, Union Carbide, General Electric and others to isolate these elements. He, also, worked hard during this period

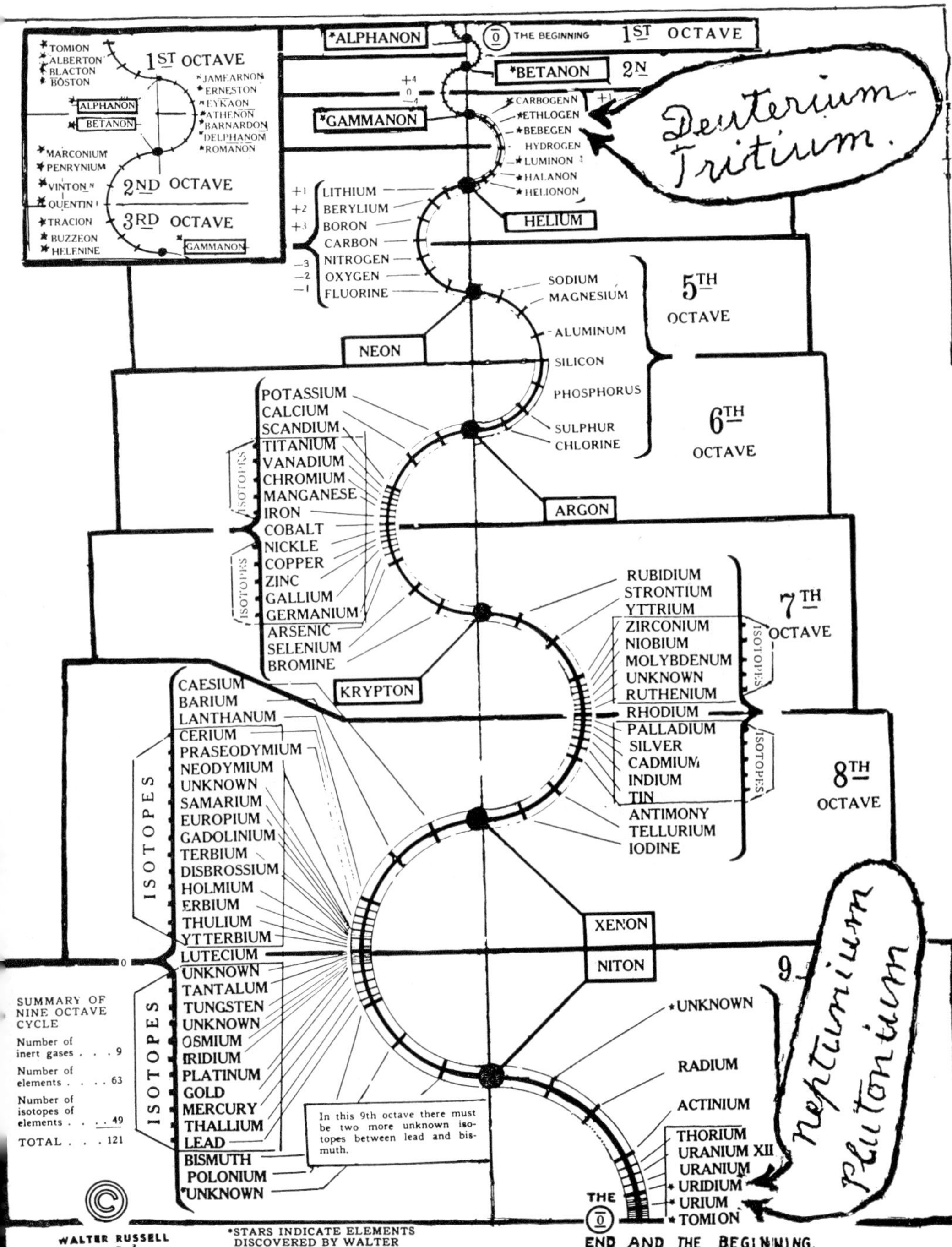

The Russell Periodic Chart of the Elements, No. 1

Periodicity is a characteristic of all phenomena of nature

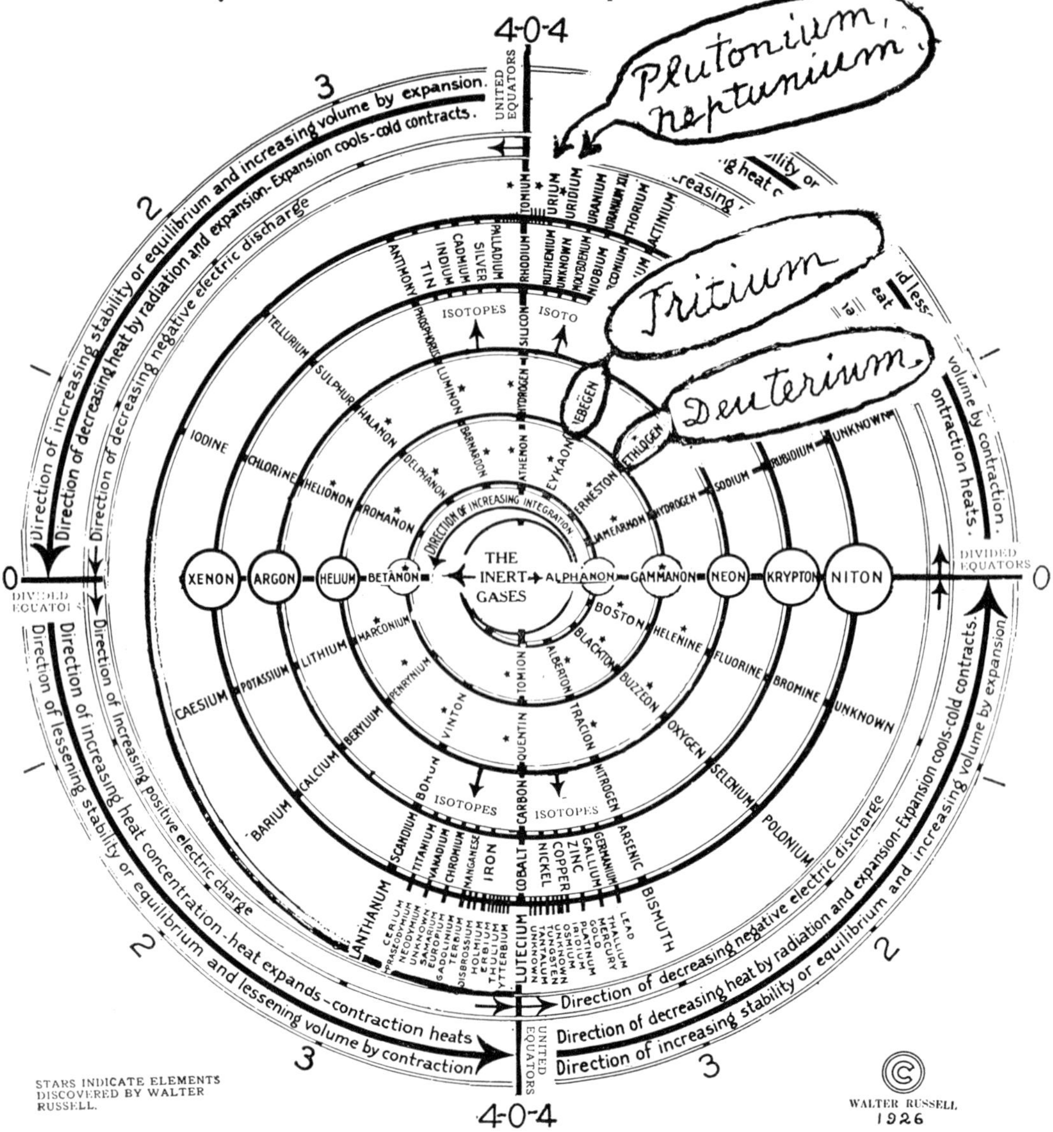

The Russell Periodic Chart of the Elements, No. 2

to urge these laboratories to produce free hydrogen as a world fuel, against their great resistance to accepting new knowledge which was outside of their pattern, and especially from one who was not an acknowledged scientist.

He foretold the great need of an inexhaustible world fuel which did not require to be transported, imported or stored. He endeavored to convince them that *free hydrogen is man's natural fuel for this coming age. It can be transmuted from the atmosphere with great simplicity if man will but change his concepts of matter and energy, which now makes transmutation impossible, to new concepts, briefly outlined in this book.* We know that this miracle could be accomplished within a few months, if men of vision and power would but open their minds to the greatest change of concept in all history.

During all of the years since 1926 he has continually tried to tell science that the true Nature of the universe is not reflected in its concepts, and that it would never be able to command matter by transmuting it until these unnatural concepts were utterly discarded and true ones accepted. The harder he tried to give his new knowledge the greater he was suppressed. A number of distinguished scientists of great vision were deeply impressed, and partially convinced, but felt that tradition was too deeply rooted to allow such a great change. These men included Thomas Edison, Dr. Harvey Rentschler, Dr. Robert Bender, Dr. H. H. Sheldon of New York University, Dr. Henry Norris Russell of Princeton University, and A. Cressy Morrison of the Union Carbide Company. Nikola Tesla was, however, fully convinced that my husband's electrical knowledge was true to Nature, but it was so different from the accepted pattern that Tesla told him that he would have to seal it in a sepulchre with instructions that it be opened in a thousand years when human intelligence had unfolded far enough to be ready to accept it.

II

This is the story of every man who is ahead of his time. It is the story of all geniuses, and especially all mystics who come into the world a thousand or more years apart.

It is the story of crucifixion, punishment, and torture of such immortals as Socrates, Jesus, Gallileo, Leonardo, Spinoza, Shakespeare, Joan of Arc, Baha'u'llah, Robert Paine, Billy Mitchell, Alexander Graham Bell, Nikola Tesla, Charles Goodyear, the Wright Brothers or General Douglas McArthur.

The ordinary average mortal may laugh, scorn, ridicule, suppress or ignore the rare few mystics and geniuses who have come into the world, but it cannot deny that every great step in human culture has come through its geniuses and every great epoch in its progress has come through its mystics.

We do not crucify on the cross these days, nor do we burn at the stake, or torture on the rack. We do the equivalent, however, in other ways, such as we did to Billy Mitchell, who tried to save us from a Pearl Harbor, or to Charles Goodyear, who gave us a new world to live in, but had to bury his child in the woods for lack of human sympathy. Thus it is that suppression of the great scientific knowledge, which my husband gave to the world, cannot hurt him, but it can hurt the whole world by keeping it enslaved to the treadmill of hard labor, war, famine and fear for another two or three thousand years—if civilization is then still existent upon the earth. I say two or three thousand years, because knowledge of the invisible controlling universe of CAUSE, can come to man only through Cosmic Conscious Illuminates, and consummate Illuminates come only once in thousands of years. It can never come to man through his senses for they only observe physical effects of Creation, without being enabled to fathom their CAUSE in the Creator. This fact is not understood by the average scientist. Albert Einstein, however, did understand it, for he once said that his greatest desire was to experience illumination into Cosmic Consciousness.

So little of that greatest of all human experiences is known to the average man, that it is difficult for him to understand how any man—especially an artist—could have such vast scientific knowledge. The fact that he did not attend any school above the primary grade, and also the fact that he acquired all of his scientific knowledge in 39 days and nights in 1921, counted against him, rather than for him. Because so little is understood by all men of this day regarding the Cosmic revelations of the mystic, the word carries with it the connotation of contempt to the average man. My husband is one of those rare few mystics to whom, from time to time in crucial periods of danger to the human race, complete knowledge of the Creator and of the processes of Creation has been given for saving humanity through giving man higher knowledge of his omnipotence. *The startling fact is that such a period of downfall of civilization is now here.*

Mankind in general does not realize that the human race is still in its intellectual infancy and that his great spiritual unfolding is still ahead. This is the reason why it is difficult for him to understand another human who has become cosmically aware of his omnipotence and omniscience.

My husband, Dr. Walter Russell. is a consummate Illuminate. God gave him an innersensory perception which reaches around the entire 360 degrees of the light spectrum. He can "see" within the atom without need of microscope or cyclotron—or within all the stars and nebulae of space without need of telescope or spectroscope. More than that he can see and know the geometry of space and the means by which the invisible universe absolutely controls the visible universe. *That means that the riddle of the universe which no man has solved, regarding the mystery of the emergence of matter from space, and of its being swallowed by space, is as clear to him as the light of the sun is clear.*

Perhaps it will be more helpful for better understanding of the miracle of this book if I briefly outline Cosmic Consciousness. Actually, my husband was unaware of it himself until the great event of 1921 when a great flash of light severed his consciousness from the seat of sensation of his body, such as that which happened to Paul in the early days, and for thirty nine days and nights he was wholly Mind, being aware of his body, and making use of it, but as though it was quite apart from him.

Dr. Richard Maurice Bucke had a very slight flash of this experience. It lasted but a few seconds but it so transformed him that he gave his life to the study of it and wrote a book about the 37 known cases in all history, only one of whom was a consummate Illuminate. A consummate Illuminate is one to whom is given all-knowledge of the Cosmos. His inner vision reaches around the entire 360 degrees of the spectrum, instead of being limited to the sense-vision of about 20 degrees. E.P. Dutton and Company have sold many editions of this book, so many people are acquainted with its nature. All people are experiencing illumination upon a small scale but do not know it. *Every flash of inspiration which man experiences is a Cosmic messsage.* It is knowledge which comes from the omnipresent Mind of God to the omnipresent Mind of man. Every genius is constantly in direct communication with God, although some are not fully aware of its meaning, as my husband was not fully aware of its meaning until his great illumination at the age of 49. He had had a number of minor illuminations from the age of 7.

My husband was a famous portrait painter at the age of 49, earning over $50,000 a year, besides earnings in architecture and the other arts, which sometimes reached $150,000. That was his status as of May 10th 1921. Up to that time he knew nothing, whatsoever, about astronomy, chemistry, electricity or the construction of matter. In fact he had no especial desire to know about science. His was the Mind of an artist. Imagine then, the amazing fact that during that timeless flash, in which he became Mind alone, he knew then all that he knows today about the Cosmos and all of God's processes, including complete knowledge of God, and God's omnipresence within man. Mind KNOWS. It does not *learn.* All-knowledge is timeless to an Illuminate.

He was not allowed to resume his union with his body, however, for thirty nine days and nights, for he had much to write down concerning the present need of man for which he was given this knowledge. During this period it was made known to him that world-agony through world-need, and population increase, versus decrease in Nature's resources, would cause man to destroy himself, and this Message was given to him to prevent that happening. He was given specific instructions about the coming wars and commanded to continue his preparation until

1946 purely as a rehearsal stage, for he was warned that man would not listen to the Message, which tore apart all old beliefs, until the world had suffered sufficiently to realize its danger.

During this 39 day period he wrote 40,000 immortal words in many fragments, without even an erasure or change of punctuation. He also made hundreds of chemical, electrical and astronomical drawings. Many of his astronomical drawings were checked for accuracy by the noted astronomer, Dr. Henry Norris Russell, and a mathematician assistant. One of them was a diagram of the whole solar system as of that day. When checked by them each planet was found to be where it should be in relation to all of the others. Other drawings of comet paths were similarly checked at Columbia and Swathmore, but during this 35 year interval many of these records—except some of the Princeton ones—have been lost. These writings and drawings are gradually being pieced together and two volumes of them have already been published as "THE MESSAGE OF THE DIVINE ILIAD" under such headings as "The Book of Love," "The Book of Rest," "The Book of Beauty," "The Book of Desire," etc. In addition to these great writings he wrote many pages of personal instruction regarding his own actions, and especially about the world-need for a new fuel which would be given to man in two stages. *The first stage to be by transmutation of the atmosphere into free hydrogen, then, generations later, by transforming solar radiation into solar generation as man's ultimate fuel. This would not only free him from dependence upon earth's resources but give him complete power to cause rains wherever he desires, on desert or meadow, and to dissipate cyclones while forming.*

During this period of severance of sensation from Consciousness my husband's family were so worried about his strange way of seeking as complete isolation as possible, and ceasing to have any interest, whatsoever, in the everyday normal affairs, that they summoned Dr. Willard H. Rogers, a well-known doctor and chemist of that day, and Charles A. Boston, President of the New York Bar Association, to examine him and furnish the necessary papers for his commitment to a mental institution. My husband had long isolated himself from his family behind closed doors of his studio and gymnasium, where he wrote and made diagrams ceaselessly, night and day. He told me it was very difficult for him to talk coherently with anyone, for he was so deeply concentrated upon that one thing that it was hard for him to "listen to little words," and give them import.

His writings and drawings were strewn everywhere, however, and when carefully examined by these two intellectuals, the doctor said: "I did not know that Walter had so deep a knowledge of chemistry but these drawings exceed my knowledge of it." The lawyer, who was a noted scholar in the classics, said: "An insane man is always illogical, but these writings are not logical but out-Whitman Walt Whitman for beauty and logic, which no man has who is not a great poet or philosopher." Thus it was that they not only refused to sign the papers but advised that something was happening which they did not understand, so counselled awaiting the time when he might descend from the great heights in which he was evidently soaring, into his former normalcy. This gradually took place, but it required a full three months to attain that normalcy.

The strange thing about his transformation to his immediate family and friends, however, was his utter unconcern about money, or a commercial adaptation of his art. From that day on his first purpose was to begin his long preparation for giving God's Message to man. He knew nothing of the language or terminology of science, yet he had to learn to speak in that language in order to tell what he knew. He had to learn the names of the elements, which he knew Cosmically as octave tones, in their wave rhythm of 0-1-2-3-4-0-4-3-2-1-0.

He knew that man's first need was to know his universe, for the Mendeleef Table of the Elements told him that man did not even suspect that there were three invisible space octaves preceding the carbon octave. He wondered why man could conceive of a universe which could build itself up to maturity of dense solidity, which carbon is, without going through its gaseous stages.

For six years Dr. Russell labored day and night to produce

the completed charts herein published and learn the language and terminology of science sufficiently to write his first epochal book, "THE UNIVERSAL ONE," together with the charts which he sent to about 800 of the world's leading scientists and universities, without the slightest evidence of their having received them, as heretofore stated.

This book is now out of print but Cambridge University had a complete photostat copy made from the New York Public Library copy, and we have heard of many other photostat copies which have been made from other originals. We have also heard that a copy has been purchased for $500.00 and much more offered for one. It will not be reprinted, for my husband was not then ready to give words to that which he knew in wordless essence.

III

The foregoing is a very brief outline to give of one who is illumined with new knowledge that could advance our world three thousand years in but ten, if those with *inner* ears will but listen to the message we shall endeavor to extend to them in this book.

From the time of his great illumination in 1921 to March 1946 my husband had worked dedicatedly alone in his tireless endeavor to awaken within man the recognition of these heretofore unknown Cosmic Laws. The most difficult thing to do is to give *new* knowledge, for man ever resists change, thinking that which he has at the moment is the "best" there is in life. Traditional teachings are so strong in man that it is easier to remember and repeat the ideas that were transcendent 1,000 years before, but which become archaic with the passing of time, than it is to think and reason in the direction of change.

My first physical contact came with my husband through that now epochal little book written by Glenn Clark entitled "THE MAN WHO TAPPED THE SECRETS OF THE UNIVERSE." A dear friend gave this to me the very month it was first released. The story of our coming together after this introduction will be told in another book but suffice it to say that that introduction—or the coming together through that little book—was as inevitable as that night should follow day. I *knew* when I read my husband's words quoted therein that the fulfillment of my own destiny had come. For the first time in my life I found in anothers words the *confirmation* of my own inner *knowing.* I recognized in him that Light which was also in me, and knew that henceforth, our two ways must be one.

It has been said that one Illuminate always recognizes another, no matter what words are used. When I read fragments of his scientific knowledge contained in that book I recognized what had also been made known to me during my own illumination —more especially what he had said about God's universe, which man terms "empty space." God had said to me *"You shall know space but never emptiness."* Glenn Clark had told how God had taken my husband up unto a high mountain and there revealed to him the rich fullness of the secrets of the universe which lie within that which was but empty space to the senses of man. The firm conviction that his thinking and mine were like the two swings of a pendulum, and the strong intuition that our message to civilization was the same message that I had been prepared for since the age of three, and the lifelong inner knowing that I was to fulfill my message with another, completely filled me with knowledge that herein was the fulfillment of my destiny. I unhesitatingly called him on the telephone from another city and was not in the least surprised to hear him say: "I recognize your voice. I have heard it within me for long years. I shall come to you at once."

When he came we immediately began to formulate plans for giving his message and mine to the world as one, for many years ahead, in spite of the fact that he was nearing 75 years of age at that time. This seems almost unbelievable, but that is actually what happened. We both knew that our destiny was one and we also knew each other as Illuminates. That was enough. From that moment our lives and all that we possessed were dedicated to the fulfillment of that which had been commanded to both of us.

God works in mysterious ways His wonders to perform. It is amazing how little man comprehends his moment to moment closeness to God. Man's individual and collective destiny are of as much a moment to moment concern of the Creator, as each brush stroke of a painter's creation is of moment to moment concern to him. One of God's ways in His giving of love to man is to transform men into geniuses and Illuminates and send them into the world to unfold the spiritual natures of man through new knowledge of man's divine Oneness with His Creator. Ever since the Dawn of Consciousness these messengers have come, and each has been given his specific work to do for man, which he must do, even though the cross awaits him for his service to man.

My husband's command is to give man a dynamic knowledge of God through knowledge of His invisible universe, and His ways and processes of creating His visible one. The purpose of my husband's long preparation for giving this knowledge is to awaken the Light within man to a knowledge of his own divinity.

My command is to give humanity a living philosophy based upon scientific knowledge of Love and God's ways and processes in relation to mate-pairs, which man has never had. Civilization can never find balance in its human relations until these principles of equal mate-hood are applied.

My specific work is to progress mankind by giving the world a mother, which the world has never had, and can never become united until it has that needed mother. God creates His ideas in mate-pairs and places one half of His every idea in its mate-half. This is a fundamental principle of Creation which is based upon the principle of balance in the giving and regiving of love.

It was made known to me that as long as this great world-home of man is man-dominated we would have wars, disunity and the lack of love which always accompanies any family home, which is man-dominated and woman suppressed. The backward countries of the world are those which suppress their women. The progress of every country is measured by its approach to man-woman mental equality. Man still remembers his killings for survival in his primate days while woman remembers her giving of life. Man-domination builds an impossible civilization built upon the man-half of God's balanced idea. It was commanded of me to balance the whole of God's idea by equalizing the mate-pairs which constitute an enduring civilization.

Knowing this we both fully understood that neither his Message of The Divine Iliad for all mankind, and my message of the motherhood of man could come fully to fruition without being both mentally fathered and mothered in accord with God's way in all things. I, therefore, fully understood the deep meaning of his words when he said to me: "At last you have come to mother my work for man and for me to father yours, for our message to man is still in the womb awaiting our unity."

This was the purpose of my long preparation which saw its first fruits in my books "GOD WILL WORK WITH YOU BUT NOT FOR YOU," "SCIENTIFIC ANSWER TO HUMAN RELATIONS" and our year's "HOME STUDY COURSE in Universal Law, Natural Science and a Living Philosophy," which my husband and I wrote together.

It is inspiring to know that the fruit of the seed we have already sown has spread around the world in the ten years which we have worked together on our mountain top at Swannanoa, as father-mother of the messages given to us to give to the world as one.

We found Swannanoa, which many thousands of people have now visited, and packed 36 tons of sculpture and art work etc., in Carnegie Hall, New York, where my husband was living when we met, and where we lived when we were first married. We moved to Virginia in the November of 1948. Here in six months we created a Shrine of Beauty which has become known as a Mountain Top Paradise. We restored the beauty of the fine old marble palace and its glorious gardens, and filled them with my husband's paintings and sculpture. Also I prepared ten large glass topped tables for the preservation and exhibition of the historical and other amazing documents concerning my husband's life, including honors from European countries, and hundreds of letters and autographs of the world's contemporary great who have either sat for portraits or visited him in his studio. One especially interesting table is devoted to controversial letters and articles for, and against his scientific concepts, which The New York Times printed over a long period.

The purpose of this collection is to demonstrate what great heights any man can reach who comprehends his omnipotence and works knowingly with the divine Mind-power which is within him, rather than with his senses. It is very gratifying to us to see so many who have spent long meditative hours here, depart with the feeling that they have found, and now comprehend, the meaning of the words: "Seek ye the kingdom of heaven within you."

The multiplication of our desire to awaken that spark of genius, which lies dormant in all men, by giving man God's Message, and demonstrating it by our great collection of the work of one man who has worked knowingly with God since the age of seven, brought new life to my husband. He has grown younger with each passing year and each year has seen the completion of some new contribution to the world. He still paints and sculps, and has completed a number of masterpieces including a beautiful bust of George Washington. Our main work, however, has been together in writing and teaching. Our basic course of study has already girdled the world and transformed many lives in all professions and varied religions, through the new Cosmic knowledge we are releasing of man's relation to man—and to God—and to Nature, which he needs to control his destiny.

Because of the resistance—which we of course quite understand—of the scientific world—we felt it better to gradually give our knowledge to those who desired and recognized it—even though it would take much much longer to reach a sufficient number that would aid in giving birth to a new era in man's unfolding.

We have used hundreds of drawings to illustrate our writings —drawings that my husband had decided to lock away in a sepulchre. I would not let him do this for I knew man's need was great TODAY and that there was not a moment to lose if we were to be ready when man's need was most dire.

For ten years we have worked indefatigably with but one thought and that one thought is to give man the knowledge that can save the whole human race from self-destruction. Our ultimate contribution is this book and the new knowledge of God's ways and processes it contains *can* give birth to the most glorious and abundant life man has ever known. The age of TRANSMUTATION is here but it cannot come into being unless man's concept of the Nature of his universe conforms to Nature's ways and processes, and until he realizes that *this is a Mind-motion-tonal universe instead of a matter-energy-substance universe.*

My husband's thirty years of seemingly hopeless effort, did not have the usual effect of building up resentments in him because he fully understood that so great a transition and reversal of traditional concept was not possible. You see he was not asking of science to make a *little* change in its concepts of the universe, he was asking it to *overthrow all of it.* He was actually trying to make science leap ahead two or three thousand years from today's concepts. He had, therefore, decided to take Nikola Tesla's advice and seal his knowledge in the Smithsonian Institute until man had sufficiently unfolded, spiritually, to be ready for it. That was what he was so actively engaged in when God directed my footsteps into this world-need for which I had also been prepared for my part in it, which was to continue it NOW, for the world-need was NOW—not a thousand years from now. It was not by chance that God sent me to my husband just two months before the time that the complete Message was to be given to mankind in the years of his greatest need, for he had reached the very limit of his physical and financial endurance.

My own Cosmic illumination which told me of the omnipotence, omniscience and omnipresence of space, instead of its presumed emptiness, was that knowledge which you will find amply revealed in this book, wherein we term it THE UNIVERSAL VACUUM. This knowledge can come to only those whose innersensory perception has reached beyond the senses into the Cosmos, as ours has reached. It is knowledge which has not yet come to earth, as you will realize in these pages.

My decision that our work must not stop now, even though it

be sealed for future man, was strongly affected by knowledge then given me that the world would increasingly suffer up to the year 1960, and that the year of 1960 would be the most crucial year in all history. I knew that there must either be an epochal transition in world-thinking, or it would become man's twelfth hour, which the last fifty years of the decline of culture has led him to. This decision, that we should continue, was also affected by finding this same knowledge in my husband's instructions, given to him in 1921. He also pictured the whole path of man from 1921 to 1960 in those instructions, which specifically told of the wars, agonies and hatreds which would bring about the suffering that was necessary to make man listen to God to find his way out of world-anguish. In those instructions he was fully informed of the fact that every empire which had grown rich by conquest of the weak would become equally poor by the arising of the weak to slay the strong.

Everything foretold to him—and to me—has already come to pass. Every rich nation is indebted in the measure of its riches, and rationing has taken the seat of plenty. For every man who has killed, ten of his own have died. The world-slave has arisen to slay his master, and to slay God's place in the human heart also, by bringing brutality into the world such as it has never known, even during its darkest ages. Thus we find the world which was pictured to us—each in his own way—a world of fear, which believes itself prosperous—a world of fuel-famine, which must again be rationed—a world so disunited, neighbor against neighbor and nation against nation, that every nation is armed to the teeth to prevent being enslaved, and every door is double locked and cities doubly policed to prevent being plundered in this greatest age of crime in all man's history.

I wrote The Scientific Answer to Human Relations to demonstrate how rich piracy-built empires had been reduced to rationing for their defiance of the law of love. Once again one nation is attempting to enslave the whole world. How many times must whole civilizations fall before finding out that such a defiance of God is impossible in Nature?

Many of the events which are now taking place today we foretold in our previous writings, but we could not reach the inner ears of those who could stem the tide of man's self-made agonies. Thus it is that my husband's message to science, and mine to humanity, must not stop. We must together give transmutation to man to give him control over motion, and together we must give him my human message of a living philosophy in order that a new civilization of human unity and abundance may be born, which will endure.

I fully know that it was God's plan that He illumine two of us as one to father and mother His Message to man in this dread hour, for God works that way in all His ways. All living things created by Him are mated pairs, which must unite as one in order to interchange and multiply. Likewise, all idea—and power—and even love—are created by Him in divided pairs, which must unite as one to interchange and multiply. *When love gives and regives equally its matings will endure, but when it takes, no power on earth can hold its matings together. That is what is the matter with the world of today.* And that is why God united us as one to multiply our power to give His Message.

As a humanist I have long been teaching in my living philosophy that balanced human matings multiply the power of both man and woman a thousand times beyond that which one could do alone, while mis-matings do—in human relations—what explosives do in mis-mated chemical relations. These things we fully knew, and it was because of our full knowledge of the unlimited power which our union would give to fulfill both the science, and the human relations need, that we withdrew into the solitude of our mountain top to sew the seed of this new knowledge and infiltrate it slowly into the hearts of those thousands of seekers for Cosmic knowledge, which this age is borning.

The knowledge which man most needs for survival and the conquest of his environment, is *scientific knowledge of his universe* and a *living philosophy which will unite all mankind in one brotherhood. Our purpose in giving this new knowledge is to consummate a marriage between science and religion which will give to the world its peace, happiness and culture, that only full awareness of the divine Light, which is within all men, can bring.*

I wrote "GOD WILL WORK WITH YOU BUT NOT FOR YOU" for the same purpose that we are writing this book, to warn mankind that he is making his own eleventh hour through *lack of knowledge of man,* and of *the nature of the universe.* Much of that book is devoted to an explanation of the physical, then the spiritual unfolding of man on his long journey from the jungle of physical struggle to his omniscient mountain-top of Cosmic Consciousness. No other book has ever told that story, for it has not yet been known. To this one purpose of making it known we have dedicated our lives.

During the time Alexis Carrel was working with my husband's humanist group he wrote "MAN THE UNKNOWN" to tell the world that of all its knowledge in many things man himself was the least known by man. In that book he said: *"The Science of Man is the most important, and least known, of all the sciences."* Edwin Markham was so imbued with that same conviction that he also expressed that same thought in these words: *"In vain do we build the city if we do not first build the man."* My husband said that my book—"GOD WILL WORK WITH YOU BUT NOT FOR YOU"—might appropriately be entitled "MAN THE KNOWN", for it is filling the purpose which Alexis Carrel and Edwin Markham told of its need.

Our "SCIENCE OF MAN HOME STUDY COURSE" might just as appropriately be filling the need for making GOD THE UNKNOWN become GOD THE KNOWN, for that is just what it is gradually doing to its world-wide students. *God has never been dynamically known for the sole reason that the nature of His universe has never yet been known.* To know one is to know *both,* for *Creation emerges from its Creator and returns to Him.* This is a clearer explanation than the text books way of saying that matter emerges from space and is swallowed up by space. Our plan of disseminating this knowledge of The Creator and His Creation was the slow one of infiltration through open minds of those increasingly more spiritually unfolded seekers who were ready for it. By this plan all men would eventually know the true nature of God and His universe. The time element, however, was too slow to prevent the certain annihilation of all organic life if radioactive fission is allowed to multiply beyond man's ability to check it. This would necessitate wholesale migrations to the arctic and antarctic regions for the few who might possibly survive this visitation of death, which man is now preparing to give to the world in the belief that he is furthering life. That is why we had to change our plan and write this book NOW.

In my own life's work as a humanist and philosopher I have touched the souls of all peoples in all lands, from Sahara desert peoples to the socially highest. I have felt fear in the universal soul and hunger in the bodies of two thirds of mankind who never knew aught but hunger. I have seen the agricultural areas of earth shrink as humans increase and deserts extend, and have pondered over the worries of science as to how the coming billions will be fed. I have listened to those who claimed that wars must continue to decimate the growing populations, and that sickened my very soul, for I have witnessed the fatalist attitude of Asians in their acceptance of famine as just something which cannot be helped. It is small wonder then, that I felt that knowledge alone could solve this problem and render famine impossible when man actually comprehended that *God's intent to give man dominion over all things did not mean physical dominion by conquest, but Mind-dominion through comprehension.*

I had heard the medical world say that synthesis of food requirements such as sugars, starches, fats, proteins and carbohydrates would be forever impossible, but my husband says that synthesis is not transmutation and all things which are created upon this earth by God, can be created by man IF and WHEN he knows *how* God creates. God does not synthesize. During those 39 days in May and June of 1921 it was made known to my husband God's full process of creation. It is very simple and can be stated in a very few words. They are as follows:

> *God divides the stillness of His omnipresent Oneness into mated pairs, and simultaneously multiplies their power to simulate His omnipotence and omniscience through fast centripetal motion.*
>
> *He then unites His mated pairs to simulate His Oneness, and simultaneously multiplies their speed of centrifugal motion until they disappear into His omnipresent stillness.*

That is its basic simplicity. It is not difficult to comprehend, and apply its complexities, when one thoroughly comprehends the above formula. The application of this knowledge is what my husband has been endeavoring to give to the world for thirty-five years, against the resistance of a world which believes that the energy which created motion *is in motion,* instead of *in its Creator,* and also believes in substance as being *real,* instead of being pure illusion created by fast motion to *simulate* substance.

This is what we are gradually infiltrating through our teachings of The Science of Man and shall briefly outline in the pages of this book. *We hope we shall cover enough of this field to convince the world that certain death will be the penalty for lack of knowledge of the Creator and His Creation, which this early stage of man's innersensory unfolding is accountable for.* Likewise, certain death by wars or famine will decimate the race if it does not learn how to rise above the necessity of using the limited fossil resources of the earth for its constantly increasing need of heat and power.

It has been said that the value of every such genius as Beethoven has been worth billions of dollars to world-culture. What a tragedy it would be if such knowledge as we are herein giving to the world in this crucial day is never made use of to give man his promised "dominion—over all the earth."

CONCLUSION

We live in a bewildering world of mysteries which man has been trying to solve for ages. Knowledge comes slowly to the unfolding intelligence of man, however. Fear is the penalty and lot of every man until knowledge makes fear impossible. In the meantime, a little knowledge and much mystery keeps fear alive in the whole human race, and out of fear for what can happen to his body comes greed, avariciousness, hatreds and enmities.

Man started that way in his jungle. Knowing nothing he feared everything, from avalanche to tornado, to wild beasts of his jungle and every other man. As knowledge began to come into the world, fear began to leave it and love became a small part of human relations.

Man knows very little about his universe and its Creator as yet, and so he still fears, hates, conquers and kills for the greed which is still a large part of the human relations of man, and he makes enemies because of his ignorance of his eternal brotherhood and Oneness with all other men.

All through this book we will gradually open new doors to knowledge of our universe, and reveal God as an ever-present reality instead of a mystery. By fully knowing the relationship of man to God, and of man to man; fear, greed, hate and enmity will depart from it and let love in. That knowledge, and that alone, will transform the human race from its present disunited fearing stage to one of unity in which the intent of the Creator and the desire of man shall see its One World ideal fulfilled.

PART I

Atomic Suicide

By

Dr. Walter and Lao Russell

PRELUDE

A wise statesman, whose name we cannot recall, once said: "If we but knew what troubles our world we might, perhaps, fix it. The quandary is that we do not know."

That is true. We do not know. We are still too near our primate days to know. We are not sufficiently unfolded as spiritual beings to recognize the fact that the cause of our world-trouble lies in that one naked fact that we are still too near our primacy to know how to build an enduring foundation.

This is a transition point for man, however, for tens of thousands of our people are far advanced in the unfolding of their spiritual natures. These are the seed of future cosmic ages. It is to these that we address these words to tell them what actually is the matter with the world. They will comprehend, even if our words have no meaning to mass-man. These only, are the ones who will transform the world in this eleventh hour of its danger, if it can be transformed. We hope it still can, but it has fallen farther in fifty years than it had gained in eight centuries.

CHAPTER I

What Is Atomic Energy?

In answering this question let it be ever remembered that God is love, and that this universe is founded upon love. Every action and its reaction in Nature must be in balance with each other in order to carry out the purposeful intent of the Creator.

Love expressed by action, means balanced fulfillment of its own law of equal giving and regiving. That is the principle upon which this purposeful universe is founded. That which purposefully fulfills the law is GOOD. It will endure and be eternally repeated with more of that which man calls GOOD. Every action of man, likewise, which is founded upon love is a fulfillment of the law, and is GOOD—and will be eternally repeated.

Our action in answering this momentous question, is founded upon love for the good of man. If it is purposeful it will bring GOOD into the world for endless repetition, which is Nature's way of working. If it is not balanced and orderly in Nature's way of normal growth of her Idea, then that moment of its departure from the GOOD will bring about its own dissolution. That is also Nature's way of dissolving that which is unrhythmic and purposeless in the fulfilling of her One Law.

. . . .

What is atomic energy? What is this great power which is now to be used for the betterment of life? That is the question which the whole world wants to know. It wants to know because there is great fear in the heart of man that it should *not* be used, but he does not know why. The only way he can know WHY radioactivity should not be used is to know WHAT radioactivity IS—and what it will DO.

We will briefly state what it is and what it can do in one short paragraph. We will then expand this one paragraph into this whole book, for the purpose of giving you full comprehension of its meaning. When you read that one paragraph, which now follows, bear in mind the fact that the supreme effort of humanity is to learn *how to live long lives in full strength,* and repeat them down through endless centuries.

Radiation is the normal death principle. Every thing in Nature dies normally by slowly radiating its heat. Radioactivity is the explosively quick death principle. Radioactivity is man's discovery of how the human race can die quickly, and not be able to propagate its kind for many long centuries.

There is your complete answer in a few words. MULTIPLIED DEATH is the new boon which this age of man believes he is giving to the furtherance of life. Naturally you do not understand it, but instinct and intuition in you are strong enough to make you *fear* it. The reason you do not understand it is because you are not yet aware of what makes things live and what makes them die.

The entire answer to this supreme question lies in knowing exactly WHAT MAKES BODIES LIVE and WHAT MAKES BODIES DIE. This cannot be briefly told. In telling it we must tell you how God constructs and dissolves matter *electrically.* If answered briefly you would not comprehend it. The greatest minds among men have been trying to discover this life principle for centuries. When we do fully answer it for you and all mankind, you will also fully comprehend your universe and all of its mysteries. The telling must be a step by step process, however, and must keep in mind the purpose of this book, which is to dynamically explain in convincing language, and undebatable postulates, that the price man will pay for the use of atomic energy is his own inevitable extinction.

Nature has a tremendously balanced rhythmic normalcy in her orderly processes of creation. It is all of it good for man and animal alike when they observe that normalcy. When they defy it, or violate it through ignorance of it, Nature does not punish them for that violation. Ignorance punishes them. The discovery of atomic energy as a means of obtaining greater heat than an orderly and normal way suitable to environment, is as much a breach of Nature's orderly processes as it would be if man violated the law of gravity by discovering that he could get to the bottom of the canyon in a much quicker way than the normal way. No man would do that because he knows better. Atomic energy for industry is as great a violation of Nature's way as a defiance of gravity would be. And it is as sure death for the violator of one as of the other.

We know that man is not intentionally defying God's law. There is but one other reason why he is doing it, which is that he does not know what he is doing. The law works just as irrevocably, and inevitably, however, against him who defies it, as it does for the one who does not know. Man admittedly does not know what he is doing. For long ages he has sought for the life principle in germ or ultimate unit of matter. He has never found it there and never will. Likewise, he has sought for the principle of death. *This he has now found, but does not know it.*

Do you know WHY you live and HOW you keep alive? Do you know WHY you die and HOW you die? Do you know "where you go when you die, or what happens to your identity after death?" No, you do not yet know these things. Mankind has not yet unfolded far enough spiritually, to know of his eternal Identity, or to know that he cannot die. *The complete argument of this story is based upon the fact that mankind does not know what either life or death is, nor that the plan that he is now making is assisting Nature in her death process by helping you to die instead of to live.* That plan is contrary to Nature's law of love. We shall tell you all you have wondered about regarding life, death and immortality, so that you will comprehend the magnitude of the crime of man against man, which man has now decided to perpetrate.

Man does not yet know his universe, nor how it became, nor how it disappears into space and reappears. There is a universal process in Creation and a motivating force for the energy which creates it. That same process applies as equally to the creation of "life," and its disappearance in "death," in your body as it does to the whole universe. Do you know what that process is? Do you know what your universe is, and God's relation to IT—and to YOU? Do you not think you could live more powerfully and wisely if you did know these things? We feel certain that when statesmen and leaders in industry know these things they will know they cannot use the death principle to aid life.

Every forthcoming page of this book will deal with the principle of life and death of the body of man, and of God's body—the universe of motion. It will, likewise, deal with the immortality of man and of his Mind-control over his orderly and normal life and death processes. It will also clearly explain that long asked question: "Where do I go when I die?"

During this explanation of the two ways of life and death, which will be clearly set forth herein, you cannot fail to see that man's illegitimate use of DEATH to benefit life will but multiply death upon this planet until not one grass blade will be left upon it. These statements present a dread picture. They seem unbelievable and impossible. It is our responsibility to prove to you that they are not only possible, but inevitable. The only way we can possibly do that is to now make those things known about life, death and immortality, which have never yet been known.

II

The real fallacy of nuclear fission for industry is that these so-called deadly poisonous gases from the radioactive elements in reactors, and in the waste products, which are encased in concrete and buried in the sea to *protect human life from their admitted danger,* are not poisons in their own environment underground, where they are serving a necessary purpose of helping to make it possible for organic life to live upon this planet. *Man makes them poisonous by removing them from their purposeful environment to place them in an abnormal environment unsuited to their normal environment.*

Animal and vegetable life are dependent upon the upper few feet of the earth's crust to live. The soil must have humus, nitrogen, carbonic gas, oxygen and water. These so-called deadly radioactive poisons are preparing the soil for oxygen dependent life to live by causing countless billions of microscopic explosions in the rock formations underground to release water and other necessities for human life and vegetation. That is all GOOD. In their proper place in Nature they are fulfilling their necessary useful purposes. They are vitally necessary where they are. They are of benefit to man when underground or distributed in rocks. *It is only when we dig them out from under the ground and condense tons of harmless rock to ounces of deadly free metal, such as uranium piles, that we make the earth uninhabitable for man.*

Think also of the hundreds of millions of years Nature has to work to decay solid rock and metal planets sufficiently to create enough decayed surface, and an atmosphere, for organic life to become possible. Mercury has no atmosphere, nor even a grass blade. Its rock surface is so hot that lead would melt there. Venus is also hot rock but a soil has begun there, and water is probably there now, but only as steam vapor, for the temperature on Venus is higher than boiling point. Our earth has had a soil and atmosphere sufficient for simple forms of life, ever since it reached its seventy or eighty million mile mark from the sun. The radioactive metals made that possible. They belong underground just as dead animal bodies belong underground. *Radioactive metals are dead and dying bodies.* That is what is not yet known of them. There are twenty-two of them which are killer metals if we take them away from their rock environment underground in Nature and make them a part of man's environment above ground.

There are many metals which will not hurt you, such as iron, copper, gold, silver and many others. This we will clarify in the next chapter in relation to Fig. 5. You can live among thousands of *living* men without being poisoned, but you cannot live upon the breath of living men without dying from their poison. You cannot in fact, live upon your own outbreathings. You would soon die from your own poison. Digging dead and dying metals from underground where their outbreathing is giving fertile soil for organic life, is like digging dead animals from where they are being transformed into living vegetable and animal life. Every farmer must put death under his ground in order to reap life above it.

The Curies procured a few grams of radium from many tons of earth. Those few grams of dead metals would spread their quick death to every cell of your body if you put them in your pocket, but they would not harm you in the slightest if you slept upon the ground above them. The radioactive metals are giving out their quick death to the rocks in which they are embedded for the purpose of expanding the rocks into the soil and water which mothers life. *It should not be dug up from the ground to expand human beings into quick death.* If you would have a good example of their purpose in life, which is beneficial to humanity, witness the great bare rock mountains of the west which are only a few million years old. Compare them with soil covered, tree covered, very much older eastern mountains, such as the Blue Ridge, White Mountains and Catskills where the soil is deep above them and waterfalls and brooks are abundant.

Water and soil are decayed and dying rock. They are, literally, dead rocks. Out of death in Nature life springs, and when life above ground is dead it gives new life to the death of earth. These are the things which mankind must know. *It knows too little about life and death, and how each one interchanges with the other to make death live and make life die.* This is so fundamental a part of the process of Nature that you would be very much more interested in her whole processes if you would but give more thought to this one. Take note, for example, that the moment a peach in the dish on your table becomes over ripe you will see fruit flies hovering over it. They were born from it, and other less complex life forms were born from it before that, just as millions of microbes and other primitive life forms are born in your living body, and more complex ones will be born from your dead body. *You do not realize that all of your body is not wholly alive all the time. Part of you is always dead or dying, up to your maturity. After that all of them are slowly dying.* The millions of microbes which await more full life in you are harmless to you while you can charge your body sufficiently to insulate your life from their death. About two pounds of your body dies every day, and the millions of dead bodies which are in that two pounds would kill your live body if you took them back into your metabolism again. *Have you ever thought of that?* Great cities must have great sanitation departments to guard against just that.

Have you ever thought that your every inbreath *charges you with power to live,* and every outbreath is death? That means that you are perpetually living and dying thirty or forty times a minute. More amazing still—if you will but give deep thought to it—is the fact that whatever lives in you is not a poison to you, but whatever of your own body which does not live *will* poison you. You breathe in *living* oxygen which comes to you from the earth and its foliage. It dies in you and you breathe out a deadly poison known as carbon dioxide. That carbon dioxide is not poison to the ground, however. It is food to the ground. The earth breathes in that food and breathes it out as oxygen, which is poison to it but life to you. *Have you ever thought of that?* Have you ever thought that the water you drink, the air you breathe, and the food you eat, comes from the dead and decaying body of this planet, and that every living body which dies and is returned to the earth, recharges the earth with renewed life? You probably have never given this a thought. We ask you to do so now, in order that you will more fully understand this next step in your understanding of life, death and immortality, in addition to beginning to know your universe.

This is your next step. We have pictured your simultaneously living-dying body. You must now know that all bodies in all the universe are the same in all respects, whether they are electrons, cells, rocks, metals, trees, men, planets or suns. All of them live and die in the same manner. All breathe in the charging breath of life and breathe out the discharging breath of death. All of them compress heat and polarize when they breathe in, and expand, cool and depolarize when they breathe out.

Why is this? It is because all bodies are made of—and by—electric waves, and every characteristic of the electric current is a life and death alternation. The entirety of Creation lives and dies in sequences. With this thought in mind consider the body of this earth in the same way that you would consider your own

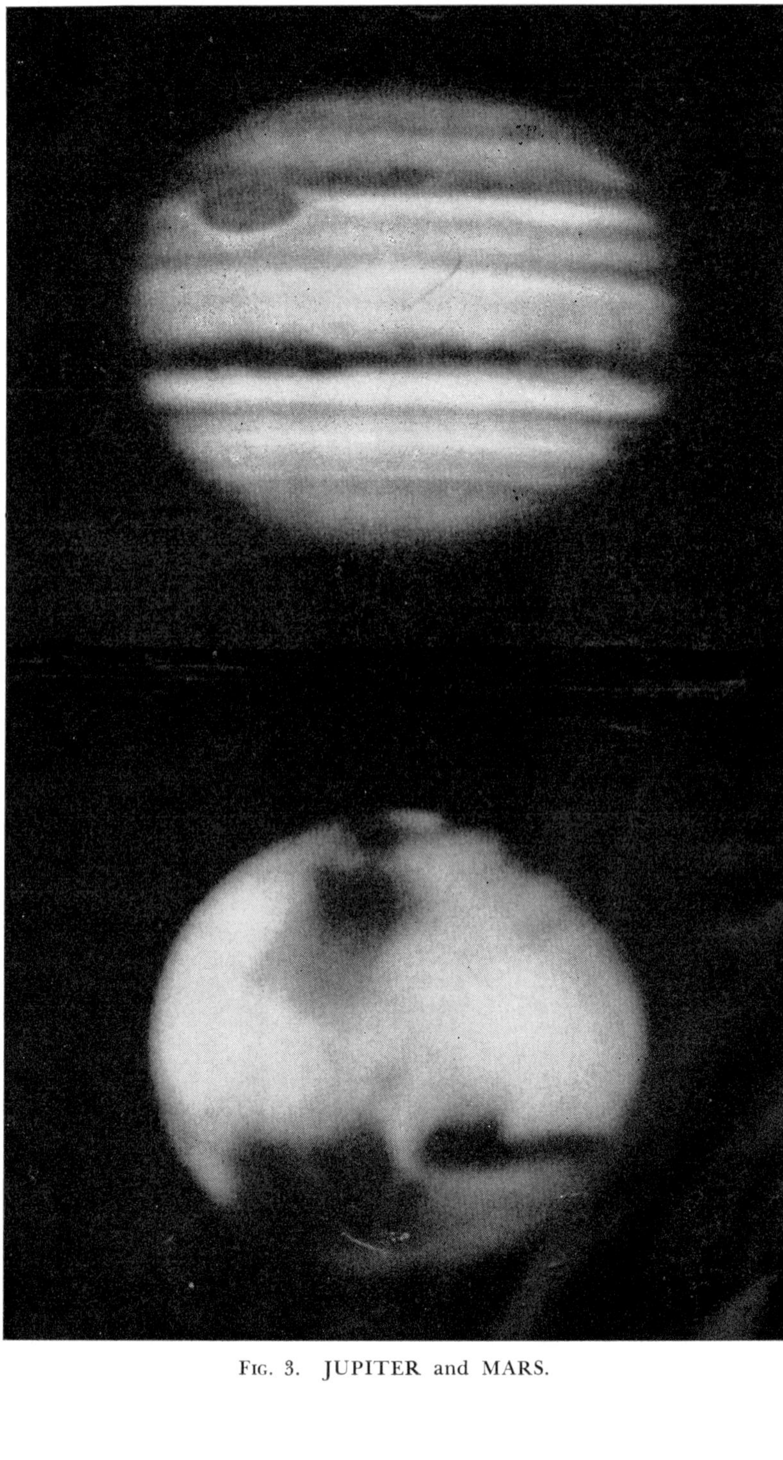

Fig. 3. JUPITER and MARS.

Fig. 4. SATURN.

body. Parts of the earth's body are dead, and other parts are dying, just as parts of your body are dead and others are dying. Just as the slight decay of an over ripe peach will not hurt you, while a fully decayed one might kill you, so, likewise, the "over ripe" chemical elements of the earth, which are not too far from carbon will not hurt you, while the farther they are beyond carbon the more deadly they become, and the more impossible it is to guard yourself from their quick death. See Fig. 5. This is the lesson of life and death which you must first know. Step by step you will know its whole process. You will then be ready to know "where you go after you die," and fully comprehend it.

Now think of the life and death of a planet as you would think of the life and death of your own body. You fully know that your own body grows to maturity—say at forty—when it is more virile than at any other age. That means that every cell in your body is more virile than that same cell could be at sixty, or eighty. At forty your body cells are true spheres and small, fast turning rings. At sixty the cells have become very much flattened at their poles as our earth is doing, and their surfaces are beginning to wrinkle with many rings, like Jupiter, just as an old face wrinkles with many lines. At eighty rings are actually thrown off at equators, like the rings of Saturn. These rings are dead cells of Saturn's still living body. An old man of ninety is still living but all of his cells are dying faster than he can replace them. This is an important fact for you to understand. Your car battery will help you comprehend the whole principle of the long slow years of dying. When it is charging while you are running it, that shows that its power to generate life is greater than its power to degenerate by discharge. The old age of anything, whether man, electron or planet, means that the galvanometer needle of its life current is showing more discharge than charge, even though it is charging *some*.

We herein print pictures of Mars, Jupiter and Saturn to illustrate three such stages of slow dying. They are all of them generating life from the pressures of the sun, but they are dying faster than they can live. See Figs. 3 and 4.

The earth is a dying cell thrown off from the sun. It cannot maintain its spherical form. It is flattening at the poles and cannot keep in balance with its system by remaining on the plane of the sun's equator. It is in the very early stages of preparing to throw off more rings such as its first one which has wound up to become a moon. The growth of deserts around its equator is the first early stage of that period. Mars has grown very much older. It still has water but oxygen dependent life is nearing its end upon it. Deserts take up a large area of it and it is more oblate than the earth.

For a good example of old age witness the wrinkled, expanded, oblate body of Jupiter. It has thrown off many rings, and is preparing to throw off more. Nature always throws off its rings in series of four. Near the surface of Jupiter you will see one series which have wound up into moons. Mercury, Venus, Earth and Mars are the last series thrown off by our sun. Jupiter has expanded to probably three or four times its original size, and is whirling itself to its centrifugal death with ever increasing acceleration. In the still more oblate Saturn you see three of its next series of rings before they have wound up around a newly established gravitative center. This study of the way planets and suns die will clearly illustrate to you how your body dies.

The above thoughts have been given you for the express purpose of building up in you a full comprehension that man is now attempting to build an environment of death upon the surface of this planet, which has for so many aeons been trying to build up an environment for life. He is creating an abnormal environment within which animal and vegetable life cannot possibly live. We desire to show you why radioactive fission, in large quantities, and oxygen cannot co-exist. These elements are fully alive while in their own environment. By taking them out from that environment they become corpses of themselves. To spread these huge ten ton reactor corpses throughout the world, and the countless billions of microscopic ones, so that they are in every breath we breathe, every bit of food we eat and in the water we drink, will eventually be like forcing you to again breathe in a gradually increasing percentage of your own dead breath.

III

From now on we must talk very much about FLAME, in respect to radioactivity and death. Flame is the ultimate of Creation. It is both maximum life and maximum death. Its radiation will beget life but its body will *consume* it. The maturity of all Creation itself is expressed in flaming suns. Suns are incandescent carbon. Carbon is frozen flame. Flame is one point of pressure beyond the maximum compression which a body will stand. Flame is the effect of the explosion which releases the compression at that ultimate point. Compression heats. *Flame is maximum heat.*

Life cannot live without heat. It is created by heat and must have it. *Radiation* gives you that heat. It is good for you. It warms you to a normalcy of temperature. It expands your body cells in perfect attunement with their compression sequences. It synchronizes with the rhythms of your breathings and your pulse beat.

Radioactivity gives you more heat than you can stand. It sends millions of alpha ray bullets into your body which accumulate there all of your life and raise your temperature, readjusting your entire metabolism until your body cells explode from their accumulated heat and expand beyond their normalcy. The slightest change in your heat abnormalcy gives you a fever. A fever means too much heat. It means expanded body cells and red blood corpuscles. Even too much heat from the sun will do that. You cannot expose yourself to its life giving rays too long else they become death giving rays. This idea you must also give much thought to, for your environment demands a proper amount of heat, which will become impossible when your present normal environment is made abnormal by countless billions of alpha and niton rays, which will bombard you constantly and show their effects in leukemia, bone cancer, deformed children and many other effects until you die by slow expansion. A "lethal dose" of radium gases will quicken that expansion.

All things die from heat expansion. That is the only way any body can die, whether it is your body, or the sun's body. *All things in Nature die normally by slow expansion. Radioactivity is multiplied expansion, which is caused by multiplied compression. It helps man die from explosively quick expansion. Flame is caused by maximum compression. Flame is the ultimate consumer of all bodies.* If you will but bear this fact in mind as we proceed you will more readily comprehend WHY and HOW this planet will soon become a barren waste if radioactivity is widely used. You will also comprehend WHY and HOW man can explode this planet into millions of fragments, or make it become a partially exploding body, which we call a nova in the heavens.

Do not for a moment think that planets and suns cannot explode. They are continually exploding. There are thousands of novas in the heavens and space is filled with countless millions of fragments of exploded planets, which we call asteroids and meteorites. These are irregular shaped lumps. Nature does not create her forms in lumps. She creates only rings and spheres which she crystalizes into geometric forms, but they must first be rings and spheres.

The one amazing thing about all matter, which is not known today, is that all matter *wants* to explode, it *wants* to die. To live is an effort. To die is effortless. *Matter is not held together from within by the attraction of gravity, as generally supposed, it is compressed together by a force exerted from the outside toward its center.* Life is hard to maintain, for that reason, and for the same reason it is easy to die because it wants to die. You must understand this fact if you wish to comprehend how radioactivity kills. It is against all modern scientific belief, because the general belief is that a material nucleus holds the atom together. *Nature* does not work that way. Atoms do not have nuclei. *Nature creates her atoms the same way that you would compress air into a tire.* It is hard work. You pump it in from the outside toward its middle. Then you put a cap on it to *imprison it.* If it is not sealed and imprisoned it would escape without effort by you. It does not need help to expand. It only needs help to be compressed.

Matter is an abnormality. *The normality of this universe is a condition of rest—an equilibrium.* Matter is not an equilibrium. It is a created condition which divides a resistant equilibrium. That division results in tremendous tensions. Tensions are not normal to the universal equilibrium. Tensions have within them a great desire for relief from tension. Decay, death, discharge, explosions and flame give matter that relief from tensions which it desires. Bear this fact in mind. Remember that the first desire of all Creation is Mind-expression by division of the universal equilibrium. Remember also that the desire to again return to the normal universal equilibrium is equally intensive. To divide an equilibrium requires work. To return to the normality of the universal equilibrium does not require work. Bear in mind, then, that it is hard to live but easy to die. Radioactivity is making it harder for bodies to live by releasing the tensions which makes them die.

Later in this book you will find that every atom is exactly like your tire. It is an electrically compressed ring with a hole in it. See Fig. 41. Nature has to seal her compressed rings to imprison that compression, just as you do, but the cap she uses to seal them is the eternal cold of space which encases every one of them with a frozen crust. That keeps the heat sealed inside, as it does to our earth. It seems strange to speak of the frozen crust of the sun, but it is frozen, nevertheless, otherwise the sun would explode.

All atoms are single units. All of these single units combine, in every octave, to make spiral atomic systems like the nebulae of space, or carbon in an octave, or a loop of force in an electric current. All atomic rings collide to unite four pairs of them into one sphere. See Fig. 29. This you will understand later in detail, but an understanding of the essence of it now will help you to comprehend the quick death principle by means of which radioactivity lets imprisoned matter escape suddenly, just as a spark lets imprisoned dynamite escape suddenly.

The special reason you should begin to understand the essence of it now is because it is against all scientific tradition and concept. It will be proven to you all through this book, however, for our whole case rests upon it. If present day belief is right in its concept that the atom is held together from the inside, then we have no case, and these writings are worthless. *If we are right in our knowledge that all matter is a powder keg which is intensely desirous of being released from its forced imprisonment, and needs only the right kind of match to explode it, you can then see that radioactive fission is that match, and these writings are priceless saviours of the race.*

It only stands to reason, therefore, that if every cell in your body wants to explode, and that radioactivity releases millions upon millions of deadly niton and alpha "matches" in our atmosphere, which will help your body cells to explode, and you breathe in enough of them, you could no more continue to live than your tire could hold its air if you open its valve. This actually happened in a laboratory where some radioactive gas escaped and a man breathed it in. It was said of him that he had a lethal dose and could not live. *Of course he could not live, nor can the whole human race and the earth's vegetation live when microscopic strontium, barium, radium and plutonium "matches" by the millions are in all the water you drink, all the food you eat, and in every breath you take.* Some of your cells will constantly explode. Strontium will give you bone cancer and make you breed terribly abnormal children. Radium and plutonium will kill you by eating your red blood cells and causing leukemia victims by the millions. We will give you the detailed reasons for this as we proceed, but we do not need to tell you what cyanide of potassium will do to you, for you know. Nor do we need to tell you what cholorine, the most deadly gas which laboratories fear, will do to you, for you know. What then can plutonium rays do to you? Dr. Ralph E. Lapp, a noted authority, tells us that they are a million times more deadly than chlorine gas, which every laboratory dreads. Also, what can it do to the planet? Its rays are said to endure for 20,000 years. It is quite possible that the earth could remain barren that long, or very much longer. Organic life would have to "evolve" again from its beginning.

That could really happen if thousands of reactors were scattered all over the world, for each reactor has a uranium pile of ten tons. That means that thousands of Frankensteins would be in our midst against which we are helpless. We could not even approach them to destroy them. Naturally that could not happen, for mankind would discover his great error long before the country could be dotted with plants, but it is more than probable that enough would be constructed, and their shields become as dangerous as they themselves are, to cause wholesale migrations to far northern latitudes for very survival. Great cities would become deserted and tens of millions would die from polluted food, water and air. The primeval life which would necessarily become imperative for the small number of survivors in the then temperate climate of the north and south poles, would set civilization back for many thousands of years, for life would be utterly impossible within a thousand miles of naked reactor plutonium piles.

One of the greatest errors in reactor construction is the use of aluminum tubes for uranium containers. Aluminum would become charged with radioactive death very quickly in comparison with silver, gold or any other metal on the blue side of the spectrum. Steel is on the red side, but its density is so much greater than aluminum that it is a better metal than aluminum for that purpose. Nickel would be better than steel, for nickel is on the blue side. The idea of constructing shields of any material at all, for protection against uranium and plutonium is only a temporary expedient. The time will come when the shields will be as dangerous as the cores. It is impossible to be otherwise. The time element may be long, but when that time comes, the greatest migration history has ever known will take place, and its road will be paved with millions of dead, largely from leukemia which will rapidly increase until it becomes a scourge.

The element of surprise which could delay the discovery of the great danger, and thus allow more plutonium piles to come into existence, is the fact that scientists are looking near the ground for fall-out dangers and other radioactive menaces. *The greatest radioactive dangers are accumulating from eight to twelve miles up. This seems to be ignored. The upper atmosphere is already charged with death dealing radioactivity for which it has not yet sent us its bill. It is slowly coming, however, and we will have to pay for it for another century, even if atomic energy plans ceased today.* That is why it is possible to have many reactors in use before discovering it is too late to stop building them. It would be impossible to destroy them for they could not be approached. Plutonium and uranium piles cannot be destroyed in any other way than by redistributing them the way Nature does. As that is utterly impossible, the only recourse left to man is to flee from them and let the centuries renormalize the earth made barren by man.

Let us quote Britain's Manchester Guardian Weekly of May, 1956, in relation to the heavy price we will soon pay from the upper atmosphere fall-out to earth: "Since records have been kept the amount of radio-strontium which has fallen out on Britain has multiplied six times. In 1948 radioactivity, to roughly two-thousandths of a gram of radium, had fallen out on each square mile of countryside. Since then the rate of deposition has increased, and now the equivalent of nearly six-thousandths of a gram of radium are settling each year. And there is plenty more of it to come. . . . The concentration of radioactivity eight miles up is some 200 times greater than it is at ground level. Between 10 per cent and 20 per cent of this is thought to settle upon the earth each year."

We know that you are wondering WHY these radioactive elements will so violently kill when others will not hurt you. We will answer this question very explicitly for you but for the present our answer is that *they will not kill if left where Nature put them to perform the beneficial service intended of them.* They are all good in their place, in fact they help to create the very water you drink and make humus for the soil, which gives you food. God's whole plan is good. When man upsets that plan he makes for himself that which he calls "bad."

Let us give you an example of our meaning. It is good and healthful for you to sit upon a great granite boulder with the sun shining warmly upon you. All of its radiations are good, for they are in normal balance with the pulsation frequencies of every cell in your body. Now if you "bombard" that 100 ton boulder and take it all apart, you will obtain fourteen ounces of uranium and two pounds of thorium. If you then sit upon that you will soon die from the explosions caused in your body cells. You can also procure other metals which will not hurt you from that boulder. You can obtain 1,000 pounds of titanium, 180 pounds of manganese, 70 pounds of chronium, 40 pounds of nickel, 10 pounds of tungsten, 5 tons of iron and 8 tons of aluminum. The reason WHY these tons of metals will not hurt you, and WHY the few ounces of thorium and uranium WILL hurt you, is the reason why it is now necessary to know the relation of LIFE and DEATH of organic bodies, to the elements of matter of which they are composed.

For this reason we again repeat *the great simple basic fact that all death is caused by expansion, and all life is caused by compression.* We also repeat the as yet unknown fact that life deeply desires to live, but has to work hard to keep sufficiently charged to maintain life, while death can fulfill its desire to die without effort. To charge a body with life takes time and much effort, but a discharge can be timeless and without effort. The flash of a short circuit is not only a good example of this fact but *is a demonstration of its proof.* If you will but submit this new thought to your reasoning you will see the naturalness, and significance of it. Death means rest, or cessation of motion. Sound is motion created with an effort, but it ceases without effort. It is an effort for you to arise and work, but it is no effort for you to cease working and rest. It is an effort to heat a rod of iron but the rod will cool without effort. You must, therefore, realize that all things will die normally and naturally without help. *That is what radiation is. Multiplied expansion means helping matter to expand quickly, and that is what radioactivity is. The use of nuclear fission, therefore, vastly multiplies the difficulty of living things to keep alive, by vastly aiding them to die.*

The mystery which we will clarify in this book is to tell how man can be master of life through knowledge, instead of letting death master him through ignorance. The first essential in knowing how to gain this control is to first realize the profundity of the basic facts above stated, and relegate all cause of normal death to normal expansion, and all quick death to abnormal expansion. Add other details to this afterward. Do not say that he died of arsenic poisoning, for there is no poison in the purposeful things of Nature, which must fulfill their purposes. *He died of abnormal expansion due to contacting a higher voltage electric current than his body could stand. That was the cause of his death. It was radioactive electrocution, not poisoning.*

This example will apply to all radioactive effects which we shall describe herein. The lethal dose of radioactive emanation which killed the laboratory worker heretofore mentioned, was not a poison, nor was it a chemical death. He died by electrocution, slow electrocution which consumed two weeks of time to slowly expand enough to kill. As our last word, before taking the next simple step in our clarification, let us remind you that this universe of motion is entirely electrical. Every effect of any nature, whatsoever, is basicly electric. Whatever happens in any way stems from the electric current. *To know just one wave unit of the electric current is to know all there is of the construction of matter, or the cause of any effect of motion, whatsoever.*

We tell our students constantly that the secret of Creation lies in the octave wave, therefore, know the wave. We, likewise, tell our students who are doctors, or chemists, lawyers, metallurgists, astronomers or inventors, that they must first be electricians. We, likewise, say the same thing to the humanist, poet or missionary. First be an electrician. Know the electric current if you wish to control people, matter, or your destiny. The chemist and musician make use of the same octave tonal scale, and the clergyman who knows its rhythms is vastly more fitted to balance human problems. We say to all men in all professions, and all walks of life, from the statesman to him who wields a hoe: "If you would know your universe of motion, your relation to it and your control over it, first thoroughly know just one cycle of an electric current and the still fulcrum from which it has its being."

THE SUPREME MYSTERY OF LIFE AND DEATH

We shall now write for you what seems like an insoluble paradox, or an impossibility of Nature, but it is a fundamental, nevertheless. Please carve it indelibly upon your consciousness now, and await its gradual comprehension as other unknown fundamentals of Nature will be gradually revealed to you herein. It is as follows:

All material bodies, organic and inorganic, live and grow by inc1easingly fast motion. They also die and decay by increasingly fast motion.

This eternal fact of Nature has always stared man in the face very conspicuously, without having been seen. Even so great a thinker and observer as Albert Einstein passed it by when he wrote his equasion of 1905, which states that fast motion *multiplies* electric potential. That fact of Nature is true, but as it is equally true that fast motion *divides* potential his whole equasion is invalid as a law, for a half truth which voids itself cannot become an eternal law.

It is not to be wondered at that this fact of Nature escaped observation by the senses, for man is so accustomed to the fact that effects disappear through slow motion, and that death is consummated by cessation of motion. The senses are not to be trusted, however. When you fully understand the above postulate you will also understand the incredible death speed of plutonium radiations which are approximately 170,000 miles per second.

The fastest speed of Nature is 186,400 miles per second. That is the *generoactive* speed in which light waves reproduce to create *living* bodies of matter. In the creation of matter the generoactive, polarizing, centripetal speed which multiplies to create solids, begins with zero at the first octave and reaches its maximum generoactive speed of 186,400 miles per second at carbon, which is the maturity of Creation. *This is the increasingly fast speed which multiplies potential. It is centripetal.* From there it begins its radioactive, depolarizing, centrifugal speed at zero and reaches 186,400 miles per second, just beyond plutonium at the amplitude point of the 9th octave wave. *This is the increasingly fast speed which divides potential to assist matter to die. It is centrifugal.* Up to nearly that point the word *fusion* has taken the place of combustion, but beyond it the speed of death is so great that the word *fission* has been coined to meet it. The speed of flame by fission causes one short circuit in about one hundred millionth of a second. Its actual speed could be computed by dividing the speed of light into the time it takes for one wave frequency.

Thus we end another lesson in unfolding your comprehension to the meaning of multiplied death before again opening new doors for still greater comprehension of the true nature of our universe.

IV

SUMMARY

New thinking is very difficult. We will make it easier to comprehend by telling it simply and in easy stages. As a first step in this direction you must know that life in a man's body is exactly the same as life in your car battery. As long as its electrical charge exceeds its discharge it lives and grows with increasing vitality. When its discharge exceeds its charge it loses its vitality and slowly dies. This is as true of a flower as of a man, or atom, or flaming sun.

We again repeat the fact that all things live and grow because of their ability to generate enough heat to charge their bodies with their normal voltage, *and no more.* Each living body has a normal temperature and pulsation frequency which must be constant in order to live and grow. Death begins with the slightest upset of that normalcy. Even one degree of extra heat beyond 98.6 will upset the entire metabolism of man. That one degree of fever is one fifth of his way to death. That is equally true of everything. *Our planet already has one degree of fever, caused by radioactivity.* That one degree has already upset the earth's metabolism enough to cause many fishes in the seas to migrate into colder waters. It is causing ice caps and glaciers to melt. *Ten more degrees of radioactive fever will make of this planet a barren waste.*

All things which die and decay do so because of their inability to continue the electric strain of compression sufficiently to preserve their normalcy. All living bodies *compress.* All dying bodies *expand.* This should not be a difficult first lesson upon life and death, for compression and expansion are the only two effects of motion and force in all this universe. Also, all things in it are eternally compressing and expanding in alternate sequences, forever. Your heart beat, your breathing, and the pulsations of every cell in your body are continually compressing and expanding in orderly ordained normal rhythms, each unit having its own normalcy of rhythmic frequencies, and each one being not only in balance with the whole, but also in balance with its normal environment.

Consider a living man, in his normal environment of temperatures and pressures, being subjected to an electric pulsation frequency of a 10,000 volt current. That tremendous abnormalcy expands every cell in his body so suddenly that they violently explode. All expanding bodies are always exploding slowly, over long periods of time. We do not use that word for slow expansion, however. We do not say that a decaying tree, which takes fifty years to go back into the ground, is exploding. If you burn it, however, the flame is a series of quick explosions which will do in two hours what Nature intended should take fifty years.

Consider your car battery in the same way. If small wires are used the charge in your battery explodes slowly. You can drive hundreds of miles because of those slow gradual explosions which gradually discharge your battery. If, however, you connect both poles with a large wire and touch the ends together you discharge the whole battery instantly. A brilliant flash of intensely hot flame appears from an environment which was cold. That is what radioactivity is. *It multiplies the frequencies of slow, invisible radiation to fast visible radioactive fusion.*

If, for another example, you wind a tungsten wire with silk, then pass a hundred thousand volt current through it in a vacuum tube, you will see a brilliant white flash of intensely white hot light which lasts for only about one millionth of a second. That light is so hot that it instantly explodes the tungsten. Nothing but helium gas remains in the tube. Also those radioactively multiplied voltage frequencies are so quick that the silk thread has not been even singed.

Fix in your mind the primary fact that *flame and radioactivity are one. Flame is quick death always.* Coal dies slowly by slow loss of heat while in the ground. It takes thousands of years to radiate itself away, while flame does it in an hour. If you multiply the frequencies of flame you are, likewise, multiplying its power to quicken the death of anything. A sulphur match will not produce a hot flame. It will burn wood *slowly* but it will not set iron on fire. The hotter flame of the acetylene torch will set iron on fire, however, but it will not set granite or concrete on fire. If you multiply the radioactive power of flame still more, by making an oxyacetylene flame, you can then electrocute granite or concrete. It will burst into flame and slowly disappear.

These great radioactive heat and multiplied frequencies are nothing, however, as compared to the intense heat which nuclear fission produces with plutonium, barium, radium and other radioactively explosive elements. They are a far cry from the flame power produced by the sulpher match. These elements have made a match for man which has produced a heat so great, and so *quickly,* that it has utterly destroyed millions of tons of rock deep down in an island and vaporized other millions of tons of sea water in *a millionth of a second.* The heat thus so timelessly generated is greater than the hottest part of our sun. *That is what we mean by describing radioactivity as quick death.*

If it could be possible to transport a hundred ton lump of the sun to our earth it would drop through it as readily as a bullet penetrates the atmosphere. Call it radioactivity, fission, fusion or what you will, its principle in Nature is the release of a frozen body from its imprisonment by releasing its binding tensions. It is the relaxing principle of Nature. It is the restoration of tense matter to its Natural state of equilibrium.

CHAPTER II

How Radioactivity Kills

I

As we sit in meditation by our fireside let our imaginations wander a bit. Here we sit in perfect comfort. Our room is warm. We can go to bed when we wish and sleep in comfort all night. We are not hungry. We have good bodies. Our temperatures are 98.6—as they should be. Our pulses are normal—as they should be. Everything about us is normal. We can get up in the morning, have breakfast, go to town in a ready waiting car, and plan our day's work and evening's relaxation without fear of interference, because everything outside our home is as normal as it is at home. That is what you want, is it not? Is not that what all civilization is striving for? Is it not a fact that human happiness, health, safety, prosperity and all things worth while depend upon the normalcy of our environment and our human relations?

Our temperature could not remain at 98.6 if this planet could be pushed five million miles nearer to the sun. Our temperature would rise. We would have a fever of several points. Naturally that could not be done, but its equivalent could be done by the use of radioactive fission. Our normal life could quite easily be made unhappy and dangerous if the town's water supply became polluted, or a hundred other things which could happen to you in your home, or to the town or nation, which would upset your normalcy.

Epidemics of typhus, polio, smallpox, and many other things have upset the normal growth of individuals and nations. Two thirds of middle Europe were wiped out by bubonic plague. A war tomorrow might draft you out of your comfortable home, your bed, your work and your relaxation. Your own desired nor-

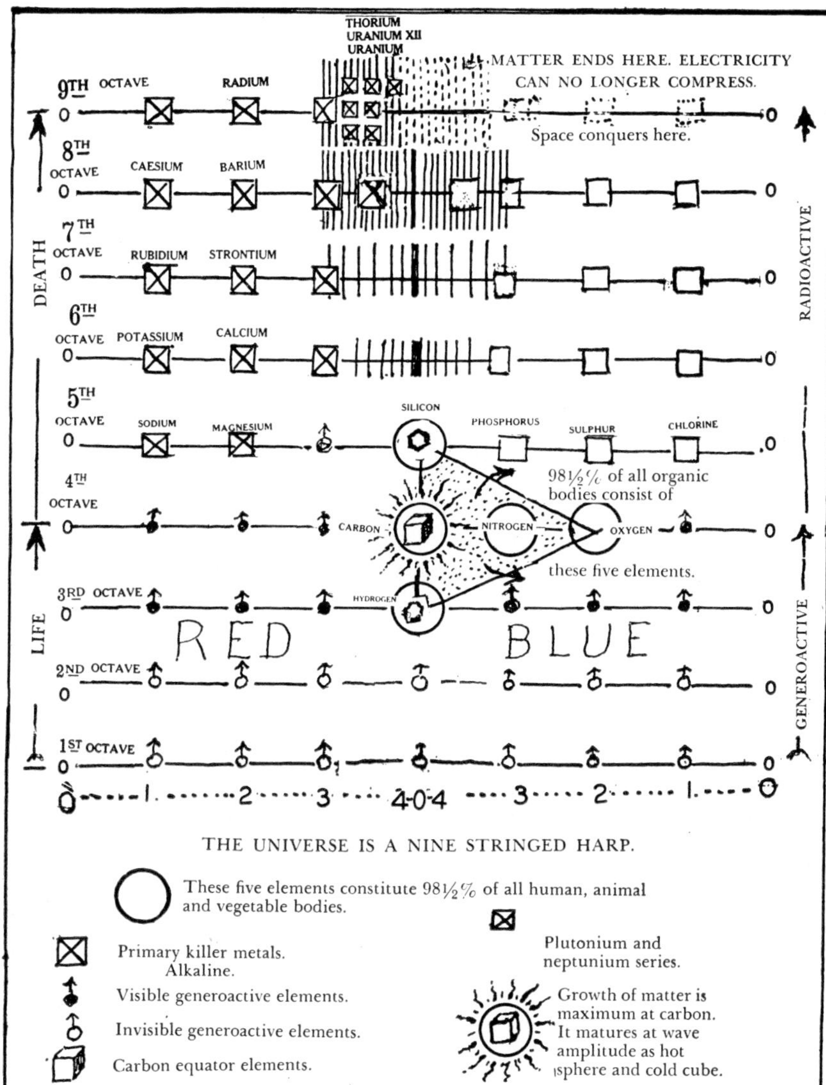

FIG. 5. The nine octaves of matter. This electric universe is a harp of nine wave strings. It produces matter by the vibration of its strings, exactly as a harp produces sounds. Different "substances" and different "sounds" are analogous. All matter consists of tonal vibrations. Matter is not substance.

malcy can be ended in a flash, and an abnormalcy take its place, which you do not desire. Instead of a happy Hungarian family and comfortable home of this minute, some are dead, and some are in prison, or exiled to Siberia in the next hour. This is true, is it not? It is quite a familiar picture to the whole world, and not one person in all the world remains untouched by the abnormalities which have upset the great orderly rhythmic growth which Nature intends for all creating things, and will give to them if they cooperate with Nature in keeping that normal rhythm of her orderly unfolding of all creating things.

Normalcy of environment is a necessity of all Creation. Eidelweis will not grow in a valley. No one would attempt to raise tropic fruits in Canada, nor could polar bears survive in Florida. If you knew that certain actions of man would so change the normalcy of your environment that it would cause an epidemic of bubonic plague, would you not rise en masse with your neighbors to prevent it? Of course you would. Bubonic plague would be a mild and transient effect as compared with what is now being planned for the human race, for many would survive it, and future centuries would forget it. *Not so with radioactive fission, however, for none would survive to forget it.* We wish to demonstrate and explain to you why oxygen dependent vegetable and animal life could not again appear upon this earth for twenty to thirty thousand years if man's plan is consummated.

II

It will be interesting and enlightening to again use your imagination regarding your comfortable and safe position in the universal house of the human race and other organic life. To aid the picturing of the position of organic life in this universe we are printing a map of the entirety of Creation for you to see it as a whole.

Figure 5 tells that whole story of Creation, and your safe house within its dangerous surroundings, which we will point out to you later. That safe house of yours in our universe is indicated by a triangle which bounds its walls of safety. You will see it right near the middle of the map. That triangle is the limit of your safe environment. Beyond it you cannot go for there are twenty-two vicious killers in the far reaches of the environment which surrounds your house, which you cannot see. You need have no fear of them, however, for Nature has chained them very securely so that they cannot hurt you while so securely imprisoned. *Man is planning to alter Nature's plan, however, and release them into your environment, even within your very house.* These killers are the invisible metallic death rays which penetrate every cell of your body with ultra-microscopic poison metal bullets at speeds of around 160,000 miles per second.

These killer metals which are soon to multiply to uncountable millions, will await you in your drinking water, which will be polluted from underground over the whole region of atomic activities, from buried atomic waste. They will await you in your food, for every grass blade will take it into the beef you eat and the milk you drink. They will also await you in every breath you take, for miles of it will accumulate in the upper atmosphere to fall upon the earth increasingly, year by year. Do not find comfort in assurance that you can be protected from it. We will consider that later, *but you can no more be protected from it than you could protect the dryness of earth from becoming wet when it rains. Neither can this planet be protected from the rise in its whole temperature, which alone would so change the earth's environment that all life would be impossible.*

III

The universe might very appropriately be likened to a harp of nine strings, with nine tones in each string. In our diagram the strings are all indicated as of the same length for simplicity. In Nature the lower string is the longest. They grow gradually shorter up to the 9th. Likewise, the tonal positions and frequencies change in each string. In every case each note on each string is one octave higher than its next lower string. You will note that each string begins with a zero and ends with a zero. These are the key notes of each octave. They are the cathodes of every electric anode. They are called inert gases, such as helium, neon, argon, krypton, etc. *They are the seed of all matter and the supreme mystery of matter.* In them is the answer to the great mystery of life and death, which shall be deferred for full explanation in Chapter 11 of this book, after preparation for its comprehension in the pages which intervene.

Perhaps you may better understand our meaning by a study of your piano keyboard. There are only seven tones there. Each eighth tone is the same as the beginning note, except it is one octave higher. The reason for that change is due entirely to pressure mathematics. Compression pressure has reached a position where the vibration frequencies have multiplied in cube ratio. If you put a book upon the top of an organ pipe you will find that its tone will jump to just one octave higher.

Like the low tones of a harp the low spectrum tones on the Cosmic harp are of so few frequencies that they cannot even be perceived by human senses until the third string is reached. Up to 1926 only one of these tones had been perceived, but at that time the Russell Charts numbered 1 and 2 in the Introduction, were published, which completed all of the octave tones of that third octave. Also, up to the issuance of the Russell Charts in 1926, the chart then in use was known as The Mendeleéf Table of the elements. In that chart hydrogen was the only element shown in that otherwise empty third octave, an impossibility in Nature which creates all units in pairs, and all octaves in four pairs. *The Russell Charts complete this deficiency.* Also, hydrogen, and its octave, were shown without an inert gas preceding it. This is as impossible in Nature as the growing of a tree without a seed is impossible. Also hydrogen was placed in the first column instead of the fourth column under carbon. Hydrogen has never been suspected as being carbon, one octave lower, as silicon is also carbon, one octave higher.

The reason that each succeeding tonal note, which constitutes the elements of matter, are one octave higher is one entirely of increased pressure. You can better understand that by compressing some air in an enclosed box or tube in which you have inserted a whistle. The more air you pump in, the higher the whistle will sound when you open its valve. If you tighten a wire the same thing will happen. The same harp string can give you many tones if you turn the tuning pins higher, or lower. By the time the first octave string has multiplied in cube ratio nine times, the speed of vibration frequencies and intensity of pressure have reached the incredible proportions of 1,073,741,824 times greater in the 9th string than the first string.

You can better imagine the tremendous explosive pressures which lie behind radium and plutonium bullets by actually seeing the pressure figures. Try it. Start with the first octave at 8. By cubing 8 it equals 64, then 512, then 4,096. The carbon octave comes next with pressure of 32,768. These are *your* normal pressures, as well as for all organic animal and vegetable life. When you cube your normal pressures you are running into danger. Nitrogen, for example, is phosphorous when cubed once, and it is arsenic when cubed twice. Nitrogen is necessary for you but phosphorous and arsenic will immediately "eat up" the nitrogen of your body and kill you by their radioactive expansion.

Oxygen is sulphur when cubed once to equal 262,144 pressures. Sulphur will kill you, however, for sulphur is the dying body of oxygen and you cannot eat your dying body. If you ignite sulphur it will consume oxygen. If you cube oxygen five times you arrive at the deadly radium-uranium-plutonium octave with its tremendous pressures of over a billion. These are the very nearly dead bodies of the elements which constitute the universal body. *The reason that these bodies poison you is the simple fact that you cannot eat the dead bodies of parts of your own living body. If we cannot survive the ptomaine poisoning of slightly decayed fish, how can we possibly survive the maximum putridity which radioactivity is?*

The deadly strontium octave reaches the tremendous pressure of 16,777,216, and the barium octave reaches 134,217,728 pressures. Underground these tremendous explosive pressures give birth to lower pressures by releasing higher ones from crystallized structures. That is the cyclic, or reincarnation process of Nature. Above ground, however, the cycle acts in reverse. Dead bodies kill living bodies instead of borning them. That is why oxygen and the free radioactive metals cannot co-exist. That is why thousands of tons of radioactive death in plutonium, strontium, thorium, radium and the other nearly dead elements used in reactor plants and discarded as waste, will gradually consume the earth's atmosphere and its oceans if not prevented from allowing the dead to remain buried, instead of resurrecting it to kill the living. We must understand that all bodies beyond carbon are dying bodies and that living bodies cannot live by consuming dead ones. There is a point of decay in all bodies, whether they are carrots, meat, fruit, oxygen or nitrogen. You do not have to be convinced that you cannot live if you consume decaying bodies of vegetables or animals which are composed of nitrogen, oxygen and carbon. Why should it be necessary to convince you that you cannot live if you consume decaying oxygen, nitrogen or carbon in their natural forms? Decay and radioactivity are one, except that radioactivity is fast decay.

Pressures in electric matter are dominated by the geometry of space wave-fields, which are based upon the cube. We recite this fact in passing, to explain the use of the word, but its further explanation would require too much space to record here. We will lightly touch upon it at the end of this book, however.

You can better understand the death dealing speed of radioactive light "bullets," which radium or plutonium can shoot at you, if you but compare the speed and power of Colonial musket bullets to the modern ones, and then multiply that by thousands of times. Or if you put sixty pounds of pressure in your tire, instead of thirty, and compare the hiss of that explosion when you open the valve, with the hiss of a thirty pound pressure, it will give you a more clear picture of what over a billion times thirty pounds would do if you could blow up a tire to such a high pressure.

IV

Let us now enter our house of safety within which our environment is all contributory to the well being of organic life. By organic life we mean oxygen dependent bodies. Within that triangle on the universal harp are only five of its tones out of its total of 121. Those five tones are carbon, silicon, oxygen, hydrogen and nitrogen. It will probably amaze you to be informed that four of these five elements constitute 98½% of your whole body composition. More amazing still, all five constitute 98½% of the upper few feet of the earth's crust upon which your body is dependent for the food it needs for survival. The basic constituent of animal bodies is carbon, while the basic constituent for the fertile life giving earth's crust, is silicon. Silicon is the first stage of death for carbon. It is, therefore, as good for purposes of earth as carbon is for purposes of life.

Our bodies need many metals but it is extremely important to realize two things about such needs. One is, that out of the many their total is only 1½% of our bodies. The other is that not even a millionth of a milligram of metal enters the body in its free state. They can be present only as mineral salts, or oxides. Iron, calcium, magnesium, sodium and many other mineral salts are necessary, but only in infinetesimal amounts. To indicate the small amounts of metals the body will accept we cite such examples as manganese, which is present in only seven hundredths of one per cent, copper one thousandth, nickel two ten thousandths, and gallium three one hundred thousandths of one per cent. Iron, calcium, iodine, sodium and other metal salts are necessary also, but in extremely small fractions.

This seems a very strange thing, for Nature creates only metals. The text books give lists of non-metals, but there are no non-metals. All stone in this universe is composed of one or more pairs of metals in union. The stone of which your grindstone is made, is the main source of aluminum, but the mate of aluminum is phosphorous. Likewise, the stony carbon is composed of boron and nitrogen in union. Nitrogen is classified as a gas, but all gases are low pressured metals. The next octave above nitrogen is phosphorous. It has been multiplied into a solid by one octave of additional pressure. Nitrogen and oxygen are good friends in the atmosphere we breathe, but phosphorous will burst into flame and consume oxygen if exposed to it, and if we breathe it we would die—yet it is but concentrated, or compressed nitrogen. If you multiply nitrogen another octave the result is arsenic, and that is obviously a metal. If you breathe arsenic vapors or take a very little of it into your body, even as a salt, it will kill you quickly—yet that too is but compressed nitrogen.

Every gas, and dense element of every octave, becomes an obvious metal when multiplied into higher octaves—and therein lies one of the reasons why organic life is possible only by keeping within the limits defined by the triangle on the map herein given.

V

We will take for the next step the as yet unknown fact that the structure of the universe, as one whole body, lives and grows to maturity, then dies, just as man, and all organic life does. Carefully note the parallel. A man matures at about forty, which is about one half of his cycle. During all of that period of time he is growing stronger and is more vital. From that moment, however, he begins to die. His generoactive maximum is reached and his radioactive period begins. He may still grow stronger in body but his cells are dying faster than he can keep them regenerated. When he is about eighty death has caught up with the power which charges bodies into life. *The WHY of that is what you need to know for your own sake, and the WHY of the death of metals is what atomic engineers must know.*

The entire birth and growth of matter matures at carbon. It matures as a flaming true sphere, any part of which will crystalize as a true cube when frozen into form. The cold cube of space and the hot sphere of matter are the consummate forms of Nature. Beyond that they cannot go. Also, they are Nature's limitations in pressures, temperatures and frequencies. Beyond the cube and the sphere they cannot go. When you understand this fundamental of Nature you will then understand WHY and HOW you and all things live, and WHY and HOW you die. Follow this explanation carefully then, and you will know. As to "WHERE you go when you die," that can wait, but that also you will know before you finish this book. We will leave this idea of cube-sphere limitation, and its reversal for a moment, but will return.

Now look at your chart and find carbon in the fourth octave

symbolized by a cube centering a sun. From that point on death begins and multiplies its power to die in the same ratio as it has multiplied its power to live. In other words, it now accumulates as much centrifugal speed to disappear into its invisible Source as it accumulated centripetal speed on the way to its destination in carbon. The only difference is that the generoactive speed of *compression is inward from the outside* until maturity in carbon, and after that its radioactive speed explodes *outward from the inside*. In each case the inward speed of 186,400 miles per second is reached at the maximum point of life in carbon, and the outward speed is reached at the amplitude of the 9th octave, just beyond the transuranium elements, which is the maximum point of universal death.

What we have just stated is one of the most significant facts of Nature. It has never been known because it has never been known that the universe lives and dies just as man lives and dies. Nor has it been known that bodies live by fast motion and also die by fast motion. All through this book you will never lose sight of that fact, for it runs through the book like a golden thread in a tapestry, as it does all through Nature.

You must fully understand that electric compression continues to generate after carbon has been reached, and death has begun to find supremacy, just as a forty year old man may still become stronger and more vital after he has begun to die. If you examine these elements beyond carbon you will find that they are all more dense and much more heavy than any element up to and including carbon. Plutonium is the heaviest of all elements. It is also the greatest of the twenty two killer metals. These killer metals are what atomic energy is intending to release into our atmosphere. We will now explain why killer metals kill and why other metals do not kill.

VI

God, the Creator is LIGHT—the invisible White Light of the undivided and unchanging Magnetic universe.

In God's Creation He limits all motion to that point in compression where invisible white light has been reached between the two visible yellows of flame in fission state. When a sun has become a true sphere its center has reached the white invisible still point in the spectrum where motion has ceased. Up to that very point the inward speed of compression has multiplied to its limit of 186,400 miles per second. At that point also the sun has reached its maximum temperature. *At that point also, there is but one center of gravity. Until then there are two.* White light is always invisible, for it is always *still.* It could not be white otherwise. Any motion, whatsoever, would be visible as yellow.

The point which we wish to emphasize by the above is that when generoactivity has created a true sphere it also has created within it a white light of gravity to center it. It has also created its maximum speed and maximum temperature. It can go no farther. The Creator has consummated His Creation. He has given all He has to give. One half of His Law of Love has been fulfilled. Up to that point The Einstein Equasion of 1905 fits perfectly. *It fully accounts for the mathematics of life, but not for death.* The other half of the Law of Love must now be fulfilled. That which has been given must be equally regiven. The balanced rhythm of this universe must not be upset. For this reason that which has been must be repeated in reverse, to void that which has been, for *neither life nor death can end or begin. They can only be repeated, and when they are repeated they do so through each other.*

You have now taken another big step in comprehending the difference between slow normal death and quick radioactive death. We must again hold this idea in suspense for a time, as it has many angles and facets which are better to be completed together, as a good diamond cutter does, instead of trying to complete one at a time.

VII

The next step for comprehension brings two words into our story which are of great import. These words are centripetal and centrifugal. Their real office in Nature has not yet been known, other than the fact that one supposedly contracts from the inside

—which is not true, and the other expands. What is not yet understood is the fact that centrifugal spiral motion, in this entire universe, is expressed where pairs of centripetal pairs meet in sex union in Nature's electric current, and begin the journey of regiving back to the Giver of motion. If this fact had been known there would have been no "expanding universe" theory. The text books say that there is no uphill flow of energy. *Centripetal motion is that uphill flow.* The cylcone vortice is that uphill flow. The winding of the universal clock spring is that uphill flow. The charging of your battery, your body, and the creative process of compression, is that uphill flow. Nature creates atom units centripetally in mated pairs. She causes atom unit pairs to collide and explode. There are not any atomic systems in Nature until mated pairs meet in sex union at wave amplitudes.

Look into the heavens. Every spinning spiral nebula is an atomic system on a stellar scale. It is made up of the dying parts of a united pair which has arrived at its centripetal maximum and must begin its dissolving journey back to its source. That dissolution is expressed by rings which wind up into spheres. These spheres throw off more rings centrifugally until all "wound up matter" has been "unwound." *Centrifugal motion is the outward direction of death. Radioactivity is centrifugal.* The power to kill, which has been generated in the twenty-two radioactive killer metals, is due to the intensively high pressure which has been compressed into them beyond their ability to hold. As a result they discharge their death rays in metal bullets for the thousands of years they consume in their dying. These metal bullets are as potent in killing you as bullets discharged from a revolver, and the principle of their projection is the same. The only difference is that radioactive bullets are so ultra microscopic that five billions of them would not make up a pinhead in size, but their expanding power within your body more than makes up for their smallness. In fact, their power to expand within your cells *is* their danger, *for expanding cells are dying cells.*

We will pause again to examine these bullets so that you will get a very clear idea of their potency, and know also that it would be as impossible for you to be protected from them as it would be impossible to protect you from feeling the effects of the sun. Let us, therefore, first examine radium expansion bullets.

VIII

We again refer you to your nine stringed Cosmic Harp. You will find radium on the 9th string. Remember that all bodies are dying from the 5th string to the 9th, although they are trying very hard to live. Look at carbon with that thought in mind. Imagine its relationship to matured life such as you would find in a virile athlete of forty. Its cube crystal form indicates its perfection of body. Now look at silicon on the 5th string. The athlete is now a virile man of fifty, but not his equal of ten years ago. Silicon and silicates can reach amplitude and the hexagon in crystal structure, but cannot reach the balanced cube.

Now look at the 6th and 7th strings. The athlete is dying with great spasmodic efforts to live. He can never reach amplitude again, however. Carbon was enabled to void its metallic quality by the union of a pair. In the 6th and 7th octaves, however, that amplitude collision is not consummated, even though five terrific efforts are made to accomplish it. The yield has been but five pairs of metals, and their inability to create a sphere with one center of gravity, instead of five *pairs* of centers, has yielded the high pressured metals known as cobalt in the 6th octave, and rhodium in the 7th.

Beyond the 7th octave the effort to live becomes greater. That tremendous effort to compress multiplies the power to die. Thirteen efforts in pairs are made in the eighth octave with the yield of a prototype of cobalt known as lutecium. You may study all of these pairs of metals by examining The Russell Periodic Chart No. 1.

In the radium octave these pairs of efforts are detectable on the red side of the spectrum, but not on the blue side, for reasons which would occupy too much space to tell here, and not sufficiently necessary to this story. As a part of it, however, take note that *all of the worst radioactive killer metals are on the red side.*

Now, as to radium. What is it? Let us examine its ancestry.

Berylium, in the 4th octave, which has a pressure intensity of 32,768, begat magnesium, which has an explosive pressure intensity of 262,144. Magnesium begat calcium, which has an explosive pressure of 2,097,152. Calcium begat strontium, at 16,777,-216 units of pressure. Strontium begat barium, at 134,217,728, and barium begat radium which has accumulated the enormous power to eject its bullets of 1,073,741,824 times that which it had at birth 9 octaves back.

Your greatest comprehension of the deadliness of the radioactive elements can come by the study of radium. We cite radium, for you can very easily visualize its deadliness by purchasing a little inexpensive instrument called a spinthariscope at any opticians for about three dollars. Within it is a fine needle which has touched a long empty, supposedly used up small vial of that very expensive radium. We cite this fact to drive home to you the potency of so inconceivable an element as this, which cannot seemingly ever be used up. It may be that several hundred thousand more spinthariscopes can still be made from that empty vial.

If you look through the lens toward a fluorescent screen you will see a sight so glorious that it could not be matched except by looking through a telescope at a star cluster. Thousands of stars seem to be exploding against that screen. What you see are the death rays of one of the most poisonous of the radioactive elements. You are seeing the luminous metallic expansion bullets which leave their metallic quality in their target to continue their expansion, and pass through, and beyond it into an inert gas named niton, then through another named zenon, then through another named krypton, and another named argon, and still another named neon until it finds its final resting place in helium. In passing through all these they have expanded them all to get back to the low pressures of the 4th octave. Further description of this principle must be deferred until the inert gases can be more fully explained in Chapter XI. We can add, however, that plutonium bullets are not content to stop at helium. They continue right through to the inert gas of the beginning of Creation in octave 1.

The wonderful, and amazing fact of this little instrument is that you could still see it as it now is for thousands of years. *That fact should answer for you the question as to the ability of man to protect you from it after thousands of tons of the still more deadly plutonium are distributed all over the country in solid ten ton piles, and not just in the scrapings of an empty milligram vial.*

An interesting story is told about radium when it was first discovered. For awhile it was thought that the life principle had at last been found. This was so sincerely believed that instruments were made to charge drinking water with these "life giving rays." Such instruments were purchased by the wealthy for as high as $1,000. Very soon, however, the tragic error was discovered before it became serious. Today, however, the danger is so well known that the number of seconds in which one is exposed to X rays, or any radioactive effect, are checked and counted to prevent too much exposure to these now known death rays.

In closing this description of radium we feel that your decision as to whether or not YOU can be protected from its dangers would be affected by the following story of the tragedy which resulted from a slight accident in a laboratory, which we are quoting from an article from Colliers. Such things are impossible to prevent. In reading it remember that it is only radium, the lesser radioactive element in that octave. That might also happen with plutonium, which is many times stronger.

> "A graphic example of how fast and far contamination can spread occurred a few years ago when someone in a Navy laboratory on Treasure Island in San Francisco Bay stepped on a glass vial containing a *barely visible* amount of radium salt. The accident was discovered late in the afternoon, and by the time decontamination crews got on the job 16 hours later, the radium had already spread throughout the San Francisco area for a radius of 20 miles.
>
> Automobiles used by students and instructors in the lab were heavily contaminated, especially the steering wheels and floor mats. Their homes were jumping with radioactivity. "It was un-

canny," recalls Lieutenant Commander Royce K. Skow, who directed much of the decontamination work. "With our instruments, we could trace the movements of the men just as though their tracks were visible. A sofa showed the outline of a student's body where he had lain down. We traced one young father from his living room to his child's crib. Two 'hot' spots showed where he had put his hands on the railing of the crib."

In a typical home, seven miles from the laboratory, a student had contaminated doorknobs, towels and water faucets. His bedspread and pillow, his slippers, his armchair, a writing desk and his pencils, his clothes—all showed radioactivity.

Since the contamination had spread outside the laboratory, where it could have been handled more effectively, drastic measures were called for. Decontamination teams ripped out carpets from a dozen homes. Automobile mats and seat covers were junked. Shoes which were only lightly affected were scrubbed again and again until instruments indicated they were clean.

The laboratory building was permeated with radioactivity and showed concentrations of radon, radioactive gas given off by radium; men entering the building wore special respirators. After a few days, crews went into the structure and burned the surface of concrete areas with scaling torches. Paint was removed. Every foot of the building was washed down. Linoleum was peeled up.

Two hundred drums of highly contaminated objects, principally carpets, clothing, linoleum and cleaning rags, were weighted with concrete and dumped into 100 fathoms of water far at sea. The building itself, a 31-room structure, could not be used for three months.

Everything connected with radioactivity is complicated by its indestructibility. A piece of contaminated newspaper cannot be burned except in a special incinerator. Even then, the ash must be carefully disposed of and gases and smokes given off in the burning must be washed or filtered. Then the poisoned water and filters must be isolated or buried."

IX

Once more we refer you to the nine octave chart. To describe radium is to describe them all, for their effects are the same. One should know them all, however. Practically all of the deadly killers are on the red side of the spectrum. You will gleam the reason for this in later chapters. They begin with three silvery white metals on the lowest of the radioactive octaves, the silicon octave. The first two are sodium and magnesium. Sodium will set water on fire and burn its oxygen out, if you throw a lump of it in water. One small pinhead of it as a free metal will kill you, but when deprived of its metal quality by union with chlorine you need it in minute quantities in your body. It is table salt. Magnesium is the familiar flashlight of photography. Naturally its radioactivity will kill you if taken into your body in its free state, yet you need a minute fraction of it as a mineral salt. Aluminum has such a density that its radioactive powers need not be feared in that octave, but in its succeeding octaves it gradually becomes more deadly than radium. Its highest form is actinium, which is an element between radium and uranium.

The sodium series, including lithium, potassium, rubidium, caesium and an unknown element one octave higher, are in the deadly class, especially because of their power to destroy oxygen by expanding it with such quick death that it bursts into flame. This effect is so little understood that a brief example will simplify it. If you touch a match to oxygen you will get a hot flame. If you compress oxygen into a liquid and then touch a match to it you will get a hotter flame, for you have multiplied its speed and expansion by multiplying its compression.

Sodium, calcium or potassium multiply the expansion of oxygen in its gaseous form and give forth the heat that liquid oxygen would give. Likewise, oxygen is multiplied in its heat-giving power if united with calcium-carbide to create an oxyacetylene flame. Consider the deadliness of potassium in this respect by the following example. You will find potassium one octave above sodium. You very freely take carbon and nitrogen into your body. They are two of your five essential elements. If you add a minute amount of potassium to them, however, you produce cyanide of potassium, a deadly quick electrocuting poison. *This is an example of what a minute quantity of radioactive matter would do if added to the essentials of our blood plasma.* That is why leukemia, birth deformities and impotence will be the forerunners of greater scourges to come.

We will again leave these thoughts in abeyance until they are more completely tied together at the end of the book.

X

We have stated before that all of the elements of matter are frozen flame. The generoactive compressive force of Nature quite easily freezes low potential explosions. Each succeeding higher potential requires greater effort to freeze it, therefore, it requires greater heat to melt it. The highest melting point of all the elements is 3,600 centigrade. This high point is reached at carbon. Silicon in the next octave reaches only 1,420 degrees and cobalt only 1,440 degrees.

The great unexplained mystery of melting points is the fact that the red side of the carbon octave reaches high melting points because of the fact that two of them are dense solids. On the blue side all three are gases and have melting points as low as 223 degrees below zero. Fluorine reaches this point while its mate on the red side, which is lithium, reaches 186 degrees above zero.

It seems strange that this octave of the highest maturity has three below zero elements on its blue side and high ones on its red side, when the dying elements have very high melting points on the blue side and less than half of those points on the red side. It is more strange still to metallurgists that many of the elements in the radioactive half of the chart are heavier and much more solid than carbon, the heaviest element on the living side of the chart. Tantalum, for example, reaches 3,400 degrees, while its close neighbors on the blue side, reach 2,900, 2,700, 2,250, 1,750 and 1,063. The answer to this is also more clearly defined later, but, very briefly, the reason is that the red side of the spectrum represents the fatherhood of Creation, which seeks the inside of forming spheres where the fires of Creation are centered, and the blue side represents its motherhood and seeks the outside to fashion bodies in her womb, and cool them into form. In the first half of the electric creating process the blue half is compressed out of the generating body, by exploding from within, while the red half gains its center, by compression exerted from the outside. In the second half the situation is reversed. All Nature is constantly reversing, and in doing so is constantly turning inside out and outside in. This conspicuous and obvious fact of Nature has not yet been sensed by observers.

SUMMARY

The foregoing has described the manner in which Nature projects motion from space and space reflects it back again. The one point which we accentuate in this whole process is the highly explosive nature of the radioactive metals. We do this for the purpose of demonstrating that this high and quick explosive power is needed in Nature's death process to assist in the decay and death of the slow dying rocky formations which first constitute the bodies of new planets, which Nature has set out from her sun crucibles to freeze. Millions of years—or perhaps billions—are consumed in the process of decaying enough of the earth's surface to create the conditions necessary for organic life. Such conditions cannot be possible nearer to the sun than seventy to eighty million miles, and ceases to be possible when a planet reaches as far out as Mars.

Water, oxygen, humus and the necessary carbohydrates of life are created by the exploding metallic bullets of the dying elements as they assist dense elements to die, but if they also project their death into less dense bodies, such as animal tissues, those lesser solids and the gases of their atmosphere, will again be assisted in their desire to expand, which is inherent in all matter.

In conclusion, therefore, we say to you that every ounce of free radioactive metal which is removed from its purposeful position of bombarding the unfertile, dry, hot rocks of Nature to yield pairs of bodies for making organic life possible on this planet, will not only clear the entire planet of all organic life but will keep it thus cleared until the many uranium piles above ground, are entirely dissolved by their own radiations. The number of thousands of years necessary for that is not predictable, but it is calculated by authorities that plutonium rays have a duration of over twenty thousand years. That is a long time. The dawn of Consciousness was but 10,000 years ago. How far back will man be, therefore, when he again appears on earth—if he ever will?

CHAPTER III

The World Voice

I

THIS CHAPTER is devoted to the world opinion and the world fear of radioactivity. It is not written to support the evidence we have given, but merely to express the world voice. We include it, also, because the world fear does not know *what* it fears. The effects are known, but the WHY of those effects are not known. That is why we must write the WHY, else the danger will be beyond remedy before the world becomes aware of it. We will cite two reasons for this statement before we quote the press—and the fears of scientists.

Our first reason is our belief that the greatest danger from the use of radioactivity is defective births and leukemia. That danger will creep upon civilization without any way of detecting it. One cannot go about with instruments to measure genetic damage, as one can do to measure the amount of strontium and other radioactivity, which is falling on the soil from year to year. We believe that sterility will be an accompanying effect, while abnormalities of living bodies will be secondary. It could not be otherwise, for genes are not basic in potency. There is something behind and underneath genes, and that is the seed. No human has ever attempted to explain the seed, therefore, it is permissible for us to say that the principle of rebirth in bodies is not yet known on earth. We know it, however, and because of that we know the danger which has no meaning to those who do not know. This mystery has to do with knowledge of the purpose of inert gases in relation to the seeds of things. We will more fully explain this mystery in Chapter XI. Right here, however, we can say that every living body refolds into its seed simultaneously with its unfolding *from* its seed. With radioactivity in general use it will not be many years before the translucent light of the inert gases, especially niton, which we described in relation to the deadly blue-white light radiations from radium, and the still more deadly blue light of uranium, will make seed-regeneration gradually impossible in either animal or vegetable species. This is the danger which will come unannounced. It is the one of which we are most apprehensive, and could cost hundreds of millions of sterile humans and more millions of defective births.

Geneticists have already begun to talk about mutations, for they understand how the seed-pattern can be altered by abnormal environment, but beyond that, to the seed itself, and its manner of refolding a dimensioned material image, such as a hundred ton oak, into a dimensionless, weightless, formless micro-pin-point inert gas recording of itself, they do not know. Nor does anyone know what the blue-white fluorescent light of inert gas from radium or from plutonium, means to the seed and to sterility. We believe, for instance, that if a reactor plant, such as the Hanford one in Washington, is placed north of New York City, as now planned, it would not be many years before the whole of that vast watershed would have to be abandoned, including many other cities near New York. To us that is as much a certainty as that a two foot high sapling will be ten feet high in a few years. It is an orderly mathematical fact of growth. In the processes of Nature it could not be otherwise. One could be forewarned of that danger before it had reached that extremity, but one cannot be forewarned of the sterility of all organic life until it has affected possibly more than half the population of the whole world in various degrees.

When radium was first used its danger was not known. No one was then forewarned of it. Because of that fact a laboratory worker named Dr. Emil H. Grubbe, was constantly exposed to it until he developed cancer burns. These did not kill him but since then he has had to undergo 90 operations because of them. He still lives at 81, but his isolated case would certainly number millions by 1970 if nuclear fission comes into general use. Does it make you feel comfortable to contemplate that fate for you, and more especially your children?

Dr. Percy Brown, a Boston X-ray specialist, had fifty operations because of various X-ray effects, including cancer. He died in 1950 after 24 years of suffering from those effects.

The well known case of accidental death which came to young Harry K. Daglian, Jr., of New London, is typical, for every known means of protection surrounded him every moment. Yet he died in 25 days because of the bomb test accident at Los Alamos, New Mexico.

Back in those days before radium danger was known, seventeen girls had been employed to paint watch dials with a radium mixed paint. One by one they all developed strange bone and blood diseases until all of them died. It has been said that such a thing could not happen now because we are forewarned. There would be thousands of cases continually happening where one could *not* be forewarned, such as the San Francisco case where someone accidentally stepped upon a very small vial of radium and exposed a whole city, which could have been wiped out because of it, had the accident not been *accidentally* discovered.

Dr. Alfred H. Sturtevant is an eminent geneticist. In 1955 he made the following statement: "The last bomb test alone probably produced more than 70 human mutations which are likely to produce large numbers of defective children in the future." Our mental and other institutions for defectives are already overflowing. The tax upon normal humans for the support of these defectives is a great burden. Such an increase in defectives as radioactivity threatens in the near future will make it imperative that all defectives be destroyed at birth. There could be no other way. Think of what our civilization would be like if it fell to such a low level as that.

Our second reason concerns the misconception of present day beliefs in relation to the ability of matter to *absorb* other matter. For a specific example, the text books describe lithium as a metal which has great power for *absorbing* water. There is no such power in all Nature. Matter does not absorb, nor contract, nor attract. Matter is compressed and *will expand and destroy, but it will not absorb.* And in that misconception lies a terrible danger, for what does actually happen is that lithium *expands oxygen and destroys it.* And so does sodium, potassium, strontium, calcium and the other silvery white killer metals, which look like silver but which you could crush between your fingers. If lithium and sodium expand oxygen in water, and your body is two thirds water, think of what it would do to your body. Always bear in mind that everything which is dying is expanding, or, conversely, expansion is the basic condition for necessary death. *Sudden expansion causes flame and great heat, for sudden expansion is an electric short-circuit.*

Another dangerous misconception is evidenced in the commonly used term: ". . . releases an extraordinary amount of energy." *Energy is not released.* What actually happens is that *potential is released, not energy—and that means expansion.* Atomic fission helps matter to explode instantly instead of decaying over long periods. "Released energy" means quick expansion, and quick expansion into the universal vacuum is the death process of any kind of material body, whether it be iron, carbon or a human blood corpuscle. That conception which gives to energy the quality of expanding or contracting, is one of the contributing factors to the deadliness of radioactivity as a power for man's use. Energy is changeless. It is the eternal vacuum of the zero universe—God's universe. To confuse the compressive power of electricity to create potential for the purpose of simulating energy, with energy itself, it extremely unfortunate.

If radioactivity ever came into general use there would be thousands of planes continually discharging ultramicroscopic particles by the billions per square mile. Thousands of ships would do, likewise, and kill every fish in the seas in time. Hundreds of reactor plants would make it impossible for you to eat any untainted food, whatsoever. Such a thing as the normal food you are accustomed to would soon be as impossible for you as it is for the cow, which is now compelled to eat contaminated grass to make into the milk you drink and the beef you eat.

Let us make more clear to you what we mean by the ability of metals like lithium, sodium, calcium or strontium to expand within you and hasten your death. A metal like iron, or gold, cannot hurt you from outside of your body because it will not expand unless you apply an intensely hot acetylene flame to it. Lithium will expand and explode in ordinary room temperature. It cannot be kept in its metallic state unless immersed in oil. If you touch an ordinary match to it a white hot flame quickly consumes it at the cost of a lot of oxygen.

Sodium will burst into flame without a match. Just throw it in water. The resultant flame does not *absorb* water, it expands the oxygen in it and sends it back into its cathode zero where all dead things go. It bursts into flame. Flame destroys. It does not *absorb*. A lesser heat than a flame expands towards death, even though it may not destroy.

In this connection the Atomic Energy Commission issued a report to the effect that the first atomic submarine, the Nautilus, had now been in operation two years without a single illness from radioactive poison taking place. Here is an inference that the atomic submarine has been proved safe. Is it? One year ago the second atomic submarine, the Sea Wolf, was launched. It is now known that its liquid sodium power system cannot be prevented from leakage and has to be removed. If a thousand such ships sailed the seas and each lost even a very little of such poison it would so utterly change the percentage of sodium in the ocean that no fish could survive it.

We have now given our two main reasons for believing that the humans who are thus dealing with death do not know the deadly nature of these killer metals, nor do they know that their *power to kill lies in their power to expand every other atom or cell of matter which they come in contact with.* It is that very power which makes them valuable and necessary in their own environment. Their ability to expand and "kill" granite, basalt, and other rock formations, gives to earth the few feet of decayed soil, atmosphere and water, necessary for organic life.

II

The following extracts from the *National Academy of Sciences* are the findings and opinions of 145 leading scientists who met for the purpose of studying this problem, under a financial grant from the Rockefeller Foundation:

"The problems of radiation fall naturally into two main classes: (1) the effects on human beings; (2) the various ways in which radiation can reach human beings through the environment.

The inheritance mechanism is by far the most sensitive to radiation of any biological system.

Any radiation which reaches the reproductive cells causes mutations (changes in the material governing heredity) that are passed on to succeeding generations.

Human gene mutations which produce observable effects are believed to be universally harmful.

Everyone is subjected to the natural background radiation which causes an unavoidable quantity of so-called spontaneous mutations. Anything that adds radiation to this naturally occurring background rate causes further mutations, and is genetically harmful.

There is no minimum amount of radiation which must be exceeded before mutations occur. Any amount, however small, that reaches the reproductive cells can cause a correspondingly small number of mutations. The more radiation, the more mutations.

The harm is cumulative. The genetic damage done by radiation builds up as the radiation is received, and depends on the total accumulated gonad dose received by people from their own conception to the conception of their last child.

So far as individuals are concerned, not all mutant genes or combinations of mutant genes are equally harmful. A few may cause very serious handicaps, many others may produce much smaller harm, or even no apparent damage.

MAY TAKE GENERATIONS

But from the point of view of the total and eventual damage to the entire population, every mutation causes roughly the same amount of harm. This is because mutant genes can only disappear when the inheritance line in which they are carried dies out. In cases of severe and obvious damage this may happen in the first generation; in other cases it may require hundreds of generations.

Thus, for the general population, and in the long run, a little radiation to a lot of people is as harmful as a lot of radiation to a few, since the total number of mutant genes can be the same in the two cases.

It is difficult to arrive at a figure showing how much genetic harm radiation can do. One measure is the amount of radiation, above the natural background, which would produce as many mutations again as occur spontaneously. It is estimated that this amounts to 30 to 80 Roentgens.

[The roentgen is a unit of radiation. To give an idea of its value, the average dental X ray deliver five roentgens to the patient's jaw, but only five thousandths of a roentgen of stray radiation to more remote parts of the body such as the gonads.]

It is also estimated that a dose of 10 roentgens to every person in the United States would cause something on the order of 5,000,000 mutant genes, which would then be a part of the population's inheritance pool. This figure is subject to considerable uncertainty.

At present the United States population is exposed to radiation from (1) the natural background, (2) medical and dental X rays, (3) fall-out from atomic weapons testing. The 30-year dose to the gonads received by the average person from each of these sources is estimated as follows:

1. Background—about 4.3 roentgens.
2. X rays and fluoroscopy—about 3 roentgens.
3. Weapons tests—if continued at the rate of the past five years would give a probable 30-year dose of about 0.1 roentgens. This figure may be off by a factor of five, that is, the possible range is from 0.02 to 0.5 roentgens. If tests were conducted at the rate of the two most active years (1953 and 1955) the 30-year dose would be about twice as great as that just stated.

At present test explosions of atomic weapons are the only significent source of radiation in the general environment, above the natural background.

Meteorologists have found no evidence that atomic explosions have changed the weather or climate. Nor do they believe that continued weapon tests, at the same rate and in the same areas as in the past, would have such an effect.

Radiation from explosions passes into the atmosphere and much of it eventually returns to the ground as "fall-out."

Fall-out divides into three classes: (1) close-in—material that comes down within a few hundred miles of the explosion and within 10 to 20 hours, (2) intermediate—material that descends in a few weeks after the explosion, (3) delayed—material that remains in the air for months or years.

Close-in fall-out from test explosions affects only restricted, uninhabited regions.

Intermediate fall-out would descend very slowly if it were pulled down only by gravity. It is mostly washed out of the air by rain and snow. It spreads over large parts of the earth, but its effect over a small area may be accentuated if there is heavy precipitation while the radioactive cloud is overhead.

Delayed fall-out is stored for long periods in the stratosphere. Meterologists know very little about the interchange of air between the stratosphere and lower layers, so they cannot predict exactly how long the material will stay up, or where it is likely to descend."

National Academy of Sciences Genetics Committee states:

"The report of the Genetics Committee of the National Academy of Sciences was unanimous and blunt: "Any radiation is genetically undesirable, since any radiation induces harmful mutations [changes]."

This complex chemical compound known as a gene cannot repair itself, and to date we know of no way of repairing it.

The injured gene will handicap some descendant, even though it may skip many generations before it does so. It may cause physical and mental handicaps to a whole line, and it will keep trying until at last it kills off the line.

Thus, the geneticists say cold-bloodedly, from the standpoint of the human race it would be better to have a few thousand humans severely radiated than to have whole populations subjected to minor radiations.

. . . .

If, as medical evidence overwhelmingly shows, man-made radiation is not good for healthy human beings, the logical questions are: Where and how might you be exposed? What can you do to protect yourself and your family?

You could be exposed through:

1. Ordinary medical X rays;
2. Eating food, drinking water or milk or breathing air that has been contaminated by fall-out from the explosion of atomic weapons tested by the United States, England and Russia;
3. Food or water contaminated by radioactive wastes from an atomic installation; . . ."

Dr. W. F. Libby says:

"Last May 2, Dr. W. F. Libby, Commissioner, United States Atomic Energy Commission, presented to the National Academy of Sciences what is probably the most authoritative public report to date on "Radioactive Strontium Fall-out."

Samples of radioactive strontium, Dr. Libby said, were detected in the snow at Admiral Byrd Bay in the Antarctic; in the waters of the Danube, the Mississippi, the Seine and the Moselle Rivers; in alfalfa in the fields of Wisconsin and Iowa, and in soil in various sections of the country.

Dr. Libby concluded his report by observing that the average content of radioactive strontium is increasing in milk supplies all over the world, and therefore more and more human beings are putting more and more radioactive strontium into their bodies.

Since growing children concentrate calcium, they are likely to absorb more of this dangerous relative of calcium into their bones than adults. And any material incorporated into their bodies during childhood will have a longer time to act."

An atomic laboratory worker dies from atomic rays:

"Philadelphia (AP)—Exposure to atomic radiation in a government laboratory between 8 and 10 years ago has taken the life of a Philadelphia physician, the city medical examiner said.

Medical Examiner Joseph W. Spelman issued his finding after an inquest into the death last July 24 of Dr. Kenneth A. Koerber, 50, of Philadelphia. Dr. Koerber had worked in the Atomic Energy Commission's Brookhaven National Laboratories, Upton, N. Y., between 1946 and 1948. He inspected laboratories to protect workers from radiation.

"We presume," Dr. Spelman said, that Dr. Koerber somehow got a dose of atomic radiation which now, 10 years later, caused

> his death. We have conclusively proved that he was subject to atomic radiation or to the inhalation or to the eating of atomic compounds. At the present time his bones contain 1,000 times the maxium safe concentration of radiation."

We quote from James Poling, in *Better Homes and Gardens,* May, 1957. The following is far from comforting, if not alarming.

> "An H-bomb explosion at our Bikini test site, March 1st, 1954, blew a scientific concept, as well as an island, into smithereens. Fallout had previously been regarded as a hazard confined to the immediate vicinity of an atomic explosion. But a recording instrument at Rongelap, 100 miles to the east of Bikini, revealed that this bomb had sprayed alarming quantities of radioactive dust over that atoll. And we learned for the first time that fallout from a multi-megaton bomb was lethal over an area of several thousand square miles. Now we know the naked horror of the bomb. It poses as a potential threat to all mankind."

The *Miami Herald* prints the following:

> DETROIT ATOM PLANT CALLED DANGEROUS
>
> "Washington—In a memo suppressed by Atomic Energy Chairman Lewis Strauss, atomic scientists have warned sternly that the "fast breeder" power reactor he approved for construction outside Detroit may "risk the health and safety of the public."
>
> Last November, a similar but smaller experimental reactor at Arco, Idaho, melted down into a hot radioactive heap that couldn't be touched for six months. Puzzled scientists still haven't found the cause.
>
> Yet Strauss, ignoring the urgent advice of his own safety experts, okayed construction of the same design atomic power plant in Detroit's populous back yard."

The eminent scientist authority upon radioactivity, Dr. Ralph Lapp, upholds our contention that the upper stratosphere will not send in its full bill for our payment until the late seventies. There is no question but what the stratosphere is accumulating death rays for destruction of earth's oxygen at an increasing rate. The freely expressed belief that "the pull of gravity" will bring the fall-out to earth is costly to human survival. A small percentage of the higher potential goes toward earth, but the larger percentage is of a potential which is lighter than the earth pressures. These seek the upper levels and the stratosphere. It is possible that enough has already collected to take toll of millions of lives, or cause millions of defective births, even if we stopped the use of radioactivity now. We print Dr. Lapp's warning in full, as issued by *The Daily Oklahoman* of July 5, 1956—and other papers:

SCIENTIST SAYS H-TESTS NEAR DANGER LEVEL

"Washington, July 4 (INS)—A leading U. S. scientist warned Wednesday that the world may already have passed the point of safety in testing hydrogen-bomb weapons.

"Dr. Ralph Lapp, citing evidence that the U. S. has exploded four H-bombs in the Pacific this year, said the world is a lot closer to the maximum safe level of radioactivity than was indicated in a recent report by the National Academy of Sciences.

"Lapp said the NAS report, issued June 13, was "misleading" when it indicated that the testing of nuclear weapons could be increased 25 to 30 times above the 1950–55 average without endangering the world's health.

THERE'S SECOND ANGLE

"He said in an interview that the NAS limit was based only on "external" radiation. He declared the safe maximum is far lower for "internal ingestion" of radioactive particles carried into the human system by air, food and water.

"The point I would emphasize," the atomic scientist said, "is that all of us already have measurable quantities of this bomb-generated radioactive material in our system.

"Although the quantities are not as yet dangerous in the cases where measurements have been taken, we now have enough information to determine definitely when the maximum safe level of radioactivity will be reached."

FOURTH BLAST HINTED

"Lapp, who has repeatedly disclosed information about the dangers of radiation which were later verified by the atomic energy commission said he believes that three super-bombs probably equal to about 10 million tons of TNT each had been set off in the current Pacific test series before Tuesday.

"The Japanese central meteorological station announced evidence of a fourth super-explosion described as "possibly" an H-bomb Tuesday morning. If the total for the series is equal to 40 million tons of TNT, this would double the average annual rate for tests conducted from 1950 through 1955.

"Lapp's warning concerned the effects of radioactive strontium, an element created when uranium atoms split. The element is chemically similar to calcium and tends to cause cancer and other ailments by concentrating in the human skeleton.

CONCLUSION CITED

"Making the much more conservative assumption that test rates would slowly accelerate and would double by 1970," he said, "I concluded that by 1962 there will be enough radioactive strontium committed to the stratosphere to produce a 100 percent 'maximum permissible amount' in every person on the planet.

"Because of a hold-up of the invisible particles in the upper air—the particles fall to earth at a rate of about 10 percent a year—this radioactive strontium would not show up in full amount in the human body until the late 1970s.

"However, by 1962 the die would have been cast irrevocably."

Instead of a slow acceleration, Lapp said, "the upward arc of bomb testing is proceeding out of control." With Russia, the U. S. and Britain conducting H-bomb experiments, he declared, the total exploded by 1960 could exceed the equivalent of one billion tons of TNT.

BATTLE NOT NEW

"Lapp has been conducting a running battle against the AEC's secret policy which prevents announcements concerning most of the tests and, even at the public detonation May 21, withheld the explosion data necessary to calculate the exact amount of radiation.

"Education about the biological effects of nuclear radiation is an absolute prerequisite if the human race is to survive," Lapp said, adding:

"I think advertisement of the probable effects of war-borne rad-

> ioactivity might also prove a valuable deterrent to the unlimited use of nuclear weapons."

Collier's was probably the first prominent magazine to vividly portray the tragic dangers of radioactivity. Two years ago it published an article about the dangers of "ATOMIC GARBAGE" by Robert De Roos, which should awaken every human to this dreadful way of exterminating the human race. We herein print copious extracts from his article. This article is the first one, ever to have been published to our knowledge, which discloses the true facts about plutonium, as being the most deadly of the killer metals. Its greater importance as a killer, lies in the fact that it loses only half of its radioactivity in 23,000 years. Its primal effect upon the human body is in its direct attack upon the bone marrow where human blood corpuscles are formed. Together with strontium, these super calciums could cause more defective skeletal births and agonizing deaths than any plague heretofore known to have hurt mankind.

When you read these extracts have this thought in mind; that radioactivity has hardly begun as yet. This fuel is not intended for temporary use, but for permanent use. Think ahead, therefore, for a hundred years. If conditions are as they now are, after but a few experimental years, what would they be in a hundred years? We believe that there will not be one man on earth in a hundred years even if so few as fifty reactors, such as Hanford and Oak Ridge, are erected.

It cannot be said that the danger to human life is not fully realized, for the cost of the waste storage tanks is $287,000 each, and the Hanford plant has already purchased $27,000,000 worth of these tanks. The fallacy of this tank plan is that the tanks themselves will disintegrate in a century or two, or become as radioactive as their contents in twenty years, thus fully releasing these deadly rays for many thousands of years of destruction. Many of these tanks have been thrown into the deep sea. We, of this generation, are quite safe from them, but other generations will pay dearly for what we are doing to them. Is not that a dreadful thought?

The latest idea for waste disposal is to lock it up in glazed bricks. Radioactive metals will destroy bricks as readily as they destroy granite and other rocky formations. It is a slow process but future generations will be the sufferers from it. There is no possible safe way of getting rid of radioactive waste beyond a few years. The deserts of the world offer the only chance. By plowing very deep furrows for hundreds of miles, and distributing all waste and free metals in very small amounts over thousands of such miles in many deserts, it may be possible to save our atmosphere. Otherwise, we will have no atmosphere in time.

One woman of the middle west, who read the following Collier's article, was heard to say: "Every pregnant woman of the future will continually fear that her baby may be a defective." What a horrible thought, especially for women of this generation, to whom such a thought never occurs.

> "Far out in the desert wasteland of eastern Washington, at the Atomic Energy Commission's gigantic Hanford plutonium works, radioactive elements surge in vast underground tanks—a pent-up sea of useless energy which is a constant worry to the scientists who unwillingly created it. This deadly broth of fission products is the garbage of the atomic age.
>
> And these highly active liquid wastes are only part of the story. Potentially dangerous atomic garbage comes in all forms: liquids, solids, gases and vapors. The ordinary defenses of man are powerless against all of them. Radioactivity is invisible and silent; it cannot be touched or tasted or smelled. And everything a radioactive element comes in contact with becomes contaminated: a wrench used in atomic installations, steel drums, a bit of wastepaper from a laboratory. Carcasses of experimental animals may contain small amounts of radioactivity; even the laundry water used to wash contaminated garments gets polluted.
>
> The story of radioactive wastes is just being understood by the public. For years it was cloaked in the secrecy which surrounds all dealings with the atom. But the dilemma posed by the wastes has been with us ever since the first self-sustaining atomic pile was activated under the bleachers at Stagg Field in Chicago in 1942, for when the physicists pulled the switch on the atomic age, they

also created something else: the world's first radioactive rubbish—the inevitable, lethal products of nuclear fission.

Ten years of production, which has seen the world-shaking atom bomb pale before the even more shattering hydrogen bomb, has left the AEC with an accumulation of millions of gallons of liquid radioactive garbage and tons of contaminated solid objects. And there's more every day.

. . . .

The problem facing the AEC—and a problem which it confidently asserts is being handled with great efficiency and success—is how to keep these new materials out of the environment—out of the air we breathe, out of our drinking water and food supplies. The garbage must be kept tightly under control because unbelievably small—often invisible—amounts can contaminate large areas.

. . . .

What makes the problem so serious is the fact that this radioactive garbage is bulking up at increasing rates—thousands of gallons every day.

And it's only the beginning. The atomic-energy industry, big as it is, is only an infant now, and today's rubbish can be handled by storing it in remote sections of the country. But under the provisions of new Congressional legislation, private manufacturers will be encouraged to develop the commercial uses of fission material. So in the not-too-distant future, disposal of wastes will become a neighborhood matter.

Some of the best engineering brains in the country are grappling with the serious question, but unless they come up with workable solutions, constantly swelling stores of waste may hamper full development of such peacetime atomic projects as generation of electric power, heating of whole communities from central atomic "furnaces," propulsion of ships and planes, and thousands of undreamed-of new ideas.

. . . .

At Hanford, where the atomic refuse represents the greatest nonmilitary concentration of radioactive elements ever known to man, the scientists rely on CC to dispose of highly active wastes. The radiation level of the hot stuff is fantastically high: the tanks the trash is buried in contain several million times the radioactivity of the world's entire commercial supply of radium.

Just one element sloughed off in making plutonium—strontium 90—is over 60 times as radioactive as radium. How dangerous is that? An engineer of the General Electric Company, which operates Hanford for the AEC, figured out what would happen in the hypothetical event that you threw the element into the Columbia River—which no one would be likely to do. After covering a sheet of paper with computations, he came up with this answer: "It would take over eight hours of the full flow of the Columbia—fifty-three billion gallons—to dilute one gram of pure strontium 90 to the point where the water would be safe enough to drink according to the AEC's permissible limits. Or putting it another way, if you dripped three grams of the element in the Columbia every day, the water of the sixth-largest river in the U. S. would be unfit to drink." (A gram is $1/28$ of an ounce.)

In addition to strontium 90, the chain reaction spews out about 40 other radioactive waste elements with half lives ranging from seconds to millions of years. A half life is the time it takes half of the atoms in a radioactive element to become disintegrated.

Storage of the really hot stuff is very expensive. A recent contract for six 1,000,000-gallon tanks was let for $1,724,000. Hanford has $27,000,000 worth of tanks with a storage capacity of 67,000,000 gallons.

. . . .

No one claims the tanks provide the solution to atomic rubbish disposal. "We are uncertain about the effects of these radioactive wastes," comments General Electric's Dr. Herbert M. Parker, director of the radiological sciences department at Hanford. "We are starting conservatively while trying to get answers which may modify the present program."

David F. Shaw, the AEC manager at Hanford, adds: "Meanwhile we keep building the tanks." At $287,000 each.

Meanwhile, most of the rubbish is stored. The only highly radioactive wastes being thrown away are those now sent out to sea to be dumped in very deep water beyond the continental shelf. A federally sponsored committee hopes this system may provide the ultimate solution to the waste-disposal problem. On the other hand, oceanographers, sanitary engineers, marine biologists and marine geologists are already concerned about what will happen when large amounts of radioactivity are thrown into the oceans.

"The sea has the same topography as the land," says Dr. Abel Wolman, of The Johns Hopkins University, a consultant to the AEC and one of the top sanitary engineers in the United States. "Big as it is, it won't accommodate everything we want to throw into it."

How winds, waves and currents will affect the dispersion or concentration of radioactive materials is under very serious study. Dr. Wolman says: "One thing that makes me a little reluctant about using the ocean is the memory of the New York garbage mess—when New York dumped its trash far at sea only to have it drift back to the Jersey beaches. And what about international control of dumping at sea?"

. . . .

At Oak Ridge, Tennessee, in the vast silences of the Appalachian Mountains, stand the tremendous gaseous diffusion plants where U235 is made by the Carbide & Carbon Chemicals Company, a division of Union Carbide & Carbon Corporation.

Oak Ridge, a 60,000-acre site, also contains the Oak Ridge National Laboratory, which produces most of the country's radioactive isotopes for experimental work.

HOW OAK RIDGE GETS RID OF ITS REFUSE

Although its waste problem does not compare with Hanford's in volume, Oak Ridge still pours 50,000 gallons of highly active refuse into its underground tanks every week. Another 5,000,000 gallons of less active rubbish also must be disposed of.

Because Oak Ridge is located in a remote region, the disposal of less active wastes is not a minor worry; but getting rid of the highly radioactive rubbish is harder. Teams of scientists, headed by Dr. Z. K. Morgan, director of the health physics division of the National Laboratory, believe they have an answer for final disposal of the dangerous, long-lived refuse at Oak Ridge, if not in all other parts of the country.

Two huge pits were bulldozed out of a hillside above White Oak Lake and large amounts of very hot wastes were poured in. The theory was that the radioactive fission products would be trapped by the fine particles of shale underlying the pits; any small amounts of hot stuff that seeped through would be diluted in the waters of the lake.

Test wells were drilled to detect movement of the rubbish

through the ground. At the end of two months, only one waste element had penetrated through to one well."

McCalls Magazine for January of 1957, had this front cover headline for a story by Pare Lorentz, "RADIOACTIVITY IS POISONING YOUR CHILDREN." We quote parts of his story which is indicative of the fear which is felt throughout the world of this threat to the human race:

"It is no secret today that a hydrogen war, no matter who starts it or who is attacked, will mean the destruction of most of the human race. One of our own high-ranking generals has stated publicly that a hydrogen attack by our forces might take the lives of three or four hundred million people—enemy *or* ally—"depending on the direction of the winds"!

What is not so well known, but what could be equally disastrous, is a possible slow degeneration of the human race caused by radioactive poisons released during times of *peace.*

. . . .

Here is a substance you can neither see, feel, hear, taste nor smell. You may be exposed to it unwittingly; you may inhale it, or absorb it by drinking polluted water or by eating contaminated food. You may not feel any immediate ill effects, yet some materials remain radioactive for years in the body, operating as so many infinitesimal but dangerous X-ray machines.

. . . .

1. *No amount of man-made radiation, external or internal, is "good" for healthy, living things. Any amount of exposure does some harm, however slight.*

. . . .

2. *Radiation is cumulative and irreversible.*

To put it very unscientifically, you have just so much radiation tolerance, and every time you are subjected to X rays or any other kind of man-made radiation, you have drawn against your total allowance. Some people have more tolerance than others, but, ideally, the quantity of total body radiation to which a human being is exposed during his lifetime should not be greater than that to which he is subjected from the sun, stars and planets and radiation arising from the minerals in the earth's surface.

3. *There is no such thing as a peaceful use of atomic energy, if by "peaceful" you mean "harmless."*
Even the by-products of a *peaceful* nuclear reactor could be used in one form or another for military purposes. Radiations inside atomic plants are dangerous. Air escaping from the plants can become contaminated. If water is used to shield or to cool the plant, it can become contaminated. The waste materials—the by-products of nuclear fission—are radioactive, some materials remaining highly dangerous for centuries.

4. *The more people exposed to radiation, the more damage to generations yet unborn.*

. . . .

For years, many groups of scientists in many parts of the world have issued appeals for a cessation of atomic explosions, on the grounds that the fall-out of radioactive materials gradually is poisoning the earth. Because of the cloud of secrecy surrounding the military facts of atomic energy and because some of these scientific groups are politically suspect, many of these appeals have been reported in the daily press as political rather than scientific discussions.

A careful study of the official reports of our own authorities, however, should convince anyone that sooner or later the atomic powers will have to stop releasing poisonous materials in the atmosphere. If they do not stop, and stop completely, they will have achieved mutual annihilation just as surely as though they had engaged in actual, all-out hydrogen warfare.

. . . .

What is more, even if all bomb tests were stopped tomorrow, it is estimated that the main part of the poisonous materials that already have been released into the upper atmosphere will not have fallen on us until 1970, and there will be a continuation of this noxious rain long after that."

The following is quoted from the *Christian Science Monitor*—as written by its Science editor, Robert C. Cowen:

"A special study group of leading American natural scientists has concluded that there is no radiation danger to human life from the present rate of atomic weapons tests. This cuts the technical ground from under the repeated demands that such tests be stopped because of the radioactive materials they release into the atmosphere.

But, at the same time, these experts warn that the full world-wide development of peaceful uses for the atom could produce more of a radiation hazard in the form of radioactive wastes than would an all-out atomic war. In fact, this "hot" waste disposal problem is growing so quickly today that their report urges its immediate and urgent consideration at the international level to determine and set up adequate safeguards.

This report by responsible natural scientists should help to bring more balanced thinking than has been evidenced in the past to bear on the vexing question of whether or not the human race is endangering its own future by releasing atomic radiations into its environment.

The report was based on extensive studies by 145 natural scientists conducted under the auspices of the National Academy of Sciences and financed by a grant from the Rockefeller Foundation. The studies were set up specifically to help resolve the world-wide controversy over radiation hazards and to provide some useful data in a field that is beset with ignorance and uncertainties."

The following is from the *Washington Daily News,* Monday, September 10, 1956, by Gene Shumate:

WE ARE BREATHING
ATOM-AGE AIR HERE

(*It's Radioactive Most of the Time*)

The air over Washington has been radioactive more than 70 per cent of the time since 1951, a Naval Research Laboratory report said today.

In 1953, it said, Washingtonians breathed pre-atomic air for only 60 days.

ATOMIC AGE

"We haven't had normal, clean air since before the atomic age —or since we started testing atomic weapons at Yucca Flats, Nev.," Dr. Herbert Friedman of the laboratory's Electron Optics Branch told The News.

Dr. Friedman is one of the authors of the report.

It said the Navy has studied the air in a series of areas since 1949—Chicago, Washington, San Diego, San Francisco, Seattle, Memphis, Puerto Rico, Panama, seven scattered Pacific isles, North Africa, and Alaska.

Of all of them, the report said, Washington and Chicago were the most radioactive.

DOSAGE IS SMALL

Dr. Friedman said the average dosage present in the air here is only two-tenths of a roentgen—far from being immediately dangerous—"but we're not sure what the accumulative effect on man will be."

It takes about 500 to 600 roentgens to kill a man outright, he said.

"There's a lot we don't know as yet about what effect breathing even slightly radioactive air over extended periods will have on us," Dr. Friedman said. "We do know that certain isotopes up there are bad actors."

The greatest and most powerful expression of fear from the use of radioactivity which has yet been made by anyone among the higher intellectuals, is the following world-broadcast appeal by Dr. Albert Schweitzer Norwegian Nobel Committee, asking that "public opinion demand an end to nuclear tests."

The New York Times, April 24, 1957.

SCHWEITZER URGES WORLD OPINION TO DEMAND END OF NUCLEAR TESTS

Nobel Winner's Plea Broadcast in 50 Lands Says Alternative Is Catastrophe for Mankind

OSLO, Norway, April 23—Dr. Albert Schweitzer has appealed to the world to end nuclear tests.

The appeal of the 82-year-old missionary surgeon, philosopher and musician, addressed to the Norwegian Nobel Prize Committee, was broadcast today in about fifty countries, including most of those in Europe. It was broadcast in Swahili from Nairobi and Japanese stations carried it.

Dr. Schweitzer's message was read in Norwegian by Unnar Jahn, chairman of the Norwegian Nobel Committee, who in 1952 bestowed the Nobel Peace Prize on Dr. Schweitzer. Translations of the text were read in English, German, French and Russian on Norwegian short-wave stations.

The impact of the warning was heightened for Norwegian listeners because the broadcast followed by fifteen minutes a report of a recent radioactive rain over Norway caused by Soviet nuclear explosions.

The initiative for the broadcast came from Dr. Schweitzer himself. The Norwegian state radio planned to send a reporter to his hospital in Lambaréné, French Equatorial Africa, where Dr. Schweitzer lives and works, to make a recording of the speech.

Dr. Schweitzer said he was too weak to read the appeal himself. It was, therefore, decided that the message would be read in translation and Dr. Schweitzer expressed the hope that it would reach the whole world.

Dr. Schweitzer said his aim was to awaken public opinion before it was too late. He warned that the human race was heading for a catastrophe if nuclear explosions were continued. This catastrophe must be prevented, he said.

"There can be no question of doing anything else, if only for the reason that we cannot take the responsibility for the consequences it might have for our descendants; they are threatened by the greatest and most terrible danger," Dr. Schweitzer said.

He emphasized that to fail to consider the importance of radioactive elements created by man and their consequences would be a folly "for which humanity would have to pay a terrible price." "We are committing this folly in thoughtlessness," he said.

PUTS IT UP TO PUBLIC

Dr. Schweitzer asked why the United States, the Soviet Union and Britain did not come to agreement to stop the tests. He believed the reason was that there was no public opinion asking for it. Japan, he added, is the only exception.

He accused "official and unofficial sources" of evading the problem when they assured that the increase in radioactivity of the air did not exceed an amount the human body could tolerate without harm.

"Even if we are not directly affected by the radioactive material in the air we are indirectly affected through that which has fallen down, is falling down and will fall down," he declared.

He also stressed that not only was the health of the present population threatened by internal radiation but also that of future generations.

"The fact is that the cells of the reproductive organs are particularly vulnerable to radiation," he said.

Dr. Schweitzer concluded his warning by appealing to public opinion in all nations to demand an agreement to stop the tests. "The end of further experiments with atomic bombs would be like the early sunrays of hope which suffering humanity is longing for," he said.

SCHWEITZER'S APPEAL TO END NUCLEAR TESTS

OSLO, Norway, April 23 (Reuters)—Following is the translation of excerpts from a letter issued by Dr. Albert Schweitzer through the Norwegian Nobel Committee, asking that public opinion demand an end to nuclear tests:

Since March 1, 1954, hydrogen bombs have been tested, by the Americans at the Pacific island of Bikini in the Marshall Group and by the Russians in Siberia.

After the explosion of a hydrogen bomb . . . something remained in the air, namely an uncalculable number of radioactive particles emitting radioactive rays. This was also the case with the uranium bombs which were dropped on Nagasaki and Hiroshima and those with which subsequent tests were made. However, because these bombs had smaller size and less effect compared with the hydrogen bombs, one hardly paid any attention to this fact.

Since radioactive rays of sufficient amount and strength have harmful effects on the human body, one started discussing if the radiation resulting from the explosions that had already taken place represented a danger which would increase with new explosions.

RACE HELD ENDANGERED

In the course of the three and a half years that have passed since then representatives of the physical and medical sciences have been studying the problem. The material collected, although far from complete, allows us to draw the conclusion that radiation resulting from the explosions which have already taken place, represents a danger to the human race, a danger not to be underrated, and that further explosions of atomic bombs will increase this danger to an alarming extent.

I raise my voice, together with those of others who have lately felt it their duty to act, in speaking and writing, as warners of the danger. My age and the sympathy that I have gained for myself through advocating the idea of reverence for life, permit me to hope that my appeal may contribute to the preparing of the way for the insight so urgently needed.

There are two kinds of atom bombs, uranium bombs and hydrogen bombs. To these two bombs has recently been added the cobalt bomb, a kind of super-atom bomb. The effect of this bomb is estimated to be many times stronger than that of hydrogen bombs having been made till now.

The explosion of an atom bomb creates an unconceivably large number of exceedingly small particles of radioactive elements.

PARTICLES HAVE LONG LIFE

Of these elements, some exist for hours, some for weeks, or months, or years, or millions of years, undergoing continuous decay. They float in the higher strata of air as clouds of radioactive dust. The heavy particles fall down first. The lighter ones will stay in the air for a longer time or come down with the rain and the snow. How long it will take before everything carried up in the air by the explosions which have taken place till now has disappeared, no one can say with any certainty. According to some estimates, this will be the case not earlier than thirty or forty years from now.

What we can state with certainty, however, is that the radioactive clouds will constantly be carried by the winds around the globe and that some of the dust, by its own weight, or by being brought down by rain, snow, mist and dew, little by little, will fall down on the hard surface of the earth, into the rivers and into the oceans.

Particularly dangerous are the elements combining long life with a relatively strong efficiency radiation. Among them strontium-90 takes the first place. It is present in very large amounts in the radioactive dust. Cobalt-60 must also be mentioned as particularly dangerous.

WATER MADE RADIOACTIVE

The radioactivity in the air, increased through these elements, will not harm us from the outside, not being strong enough to penetrate the skin. But the danger which has to be stressed above all the others is the one which arises from our drinking radioactive water and our eating radioactive food as a consequence of the increased radioactivity in the air.

Following the explosions on Bikini and Siberia rain falling over Japan was, from time to time, been so radioactive that the water from it cannot be drunk. And not only there: reports of radioactive rainfall are coming from all parts of the world where analyses have recently been made. In several places, the water has proved to be so radioactive that it was unfit for drinking.

Wherever radioactive rainwater is found the soil is also radioactive—and in a higher degree. The soil is more radioactive not only by the downpour, but also from radioactive dust falling on it. And with the soil the vegetation will also have become radioactive.

The radioactive elements deposited in the soil pass into the plants where they are stored. This is of importance, for as a result of this process it may be the case that we are threatened by a considerable amount of radioactive elements.

The radioactive elements in grass, when eaten by animals whose meat is used for food, will be absorbed and stored in our bodies.

What this storing of radioactive material implies is clearly demonstrated by the observations made when, at one occasion, the radioactivity of the Columbia River in North America was analyzed. The radioactivity was caused by the atomic plants at Hanford, which produce atomic energy for industrial purposes, and which empty their waste water into the river.

FINDS PROBLEM EVADED

The radioactivity of the river water was insignificant. But the radioactivity of the river plankton was 2,000 times higher, that of the ducks eating the plankton 40,000 times higher, that of the fish 150,000 times higher. In young swallows fed on insects caught by their parents in the river, the radioactivity was 500,000 times higher and in the egg yolks of water birds more than 1,000,000 times higher.

From official and unofficial sources we have been assured, time and time again, that the increase in radioactivity of the air does not exceed the amount which the human body can tolerate without any harmful effects. This is just evading the problem.

Even if not directly affected by the radioactive material in the air, we are indirectly affected through that which has fallen down, is falling down, and will fall down. We are absorbing this through radioactive drinking water and through animal and vegetable foodstuffs, to the same extent as radioactive elements are stored in the vegetation of the region in which we live. Unfortunately for us, nature hoards what is falling down from the air.

None of the radioactivity of the air, brought into existence by the exploding of atom bombs is so unimportant that it may not, in the long run, become a danger to us through increasing the amount of radioactivity stored in our bodies.

What are the diseases caused by internal radiation? The same diseases that are known to be caused by external radiation.

They are mainly serious blood diseases. If the cells in the bone marrow are damaged by radiation they will produce too few or abnormal, degenerating blood corpuscles. Both cases lead to blood diseases and, most often, to death. These were the diseases that killed the victims of X-rays and radium rays.

It was one of these diseases that attacked the Japanese fishermen who were surprised in their vessel by radioactive ashes falling down 240 miles from Bikini after the explosion of an hydrogen bomb. With one exception, they were all saved, being strong and relatively mildly affected, through continuous blood transfusions.

DAMAGE TO DESCENDANTS

In the cases cited, the radiation came from the outside. It is unfortunately very probable that internal radiation affecting the bone marrow and lasting for years will have the same effect, particularly since the radiation goes from the bone tissue to the bone marrow.

Not our own health only is threatened by internal radiation, but also that of our descendants. The fact is that the cells of the reproductive organs are particularly vulnerable to radiation. To the profound damage of these cells corresponds a profound damage to our descendants.

To find out how the existing radioactive radiation has affected posterity, comparative studies have been made between the descendants of doctors who have been using X-ray apparatus for years and those of doctors who have not. Among the descendants of radiologists, a percentage of still births of 1.403 was found, while the percentage among the nonradiologists was 1.222.

In the first group, 6.01 per cent of the children had congenital defects, while only 4.82 per cent in the second.

It must be remembered that even the weakest of internal radiation can have harmful effects on our descendants.

The total effect of the damage done to descendants of ancestors who have been exposed to radioactive rays will not, in accordance with the laws of genetics, be apparent in the generations coming immediately after us. The full effects will appear only 100 or 200 years later.

We are forced to regard every increase in the existing danger through further creation of radioactive elements by atom bomb explosions as a catastrophe for the human race, a catastrophe that must be prevented under every circumstance.

There can be no question of doing anything else, if only for the reason that we cannot take the responsibility for the consequences it might have for our descendants.

They are threatened by the greatest and most terrible danger.

That radioactive elements created by us are found in nature is an astounding event in the history of the earth. And of the human race. To fail to consider its importance and its consequences would be a folly for which humanity would have to pay a terrible price. When public opinion has been created in the countries concerned and among all nations, an opinion informed of the dangers involved in going on with the tests and led by the reason which this information imposes, then the statesmen may reach an agreement to stop the experiments.

A public opinion of this kind stands in no need of plebiscites or of forming of committees to express itself. It works through just being there.

The end of further experiments with atom bombs would be like the early sun-rays of hope which suffering humanity is longing for.

New York Times, April 21st. 1957.

EUROPE'S NUCLEAR FEAR

An Analysis of Changed Sentiment Toward Danger of Atomic Warfare

PARIS, April 20—Europe is growing alarmed about nuclear bombs. The mood today has markedly changed from that of the late decade, when the United States monopoly or predominance in nuclear power was widely regarded as assuring peace and the security of this Continent.

Nuclear weapons are no longer considered a beneficent invention. At the very moment when the United States officially promised nuclear arms to its European allies, eighteen West German physicists joined in declaring they opposed the use of such arms by their country and would do nothing to further nuclear armament. They said current tactical nuclear weapons were as destructive as the original atomic bomb dropped on Hiroshima and that their acquisition by West Germany might increase its danger.

Europeans are becoming convinced that nuclear weapons are a menace even if there never is a nuclear war. A committee of the Atomic Scientists Association in Britain has reported that as a result of the hydrogen bombs exploded so far one person in 50,000 is likely to get cancer in the next few decades from the strontium entering his bones. This is an unproved estimate but represents a scientific judgment based on the assumption that the likelihood of cancer is proportionate to the radioactivity in the bones.

Pending more conclusive evidence and fuller agreement among experts, many Europeans favor suspension of the bomb tests that produce radioactive "fall-out" of greater or lesser malignancy. British statesmen find it more difficult to defend their nuclear bomb tests, which are only beginning.

Soviet bomb tests have led to disturbing atmospheric results in Japan. After measuring the radioactive fall-out this week, Japanese scientists warned of the danger and opinion grew in favor of a ban on all nuclear tests by all nations. An official Japanese protest to Moscow brought the reply that the Soviet Union was willing to suspend the tests if the Western powers agreed.

Thus a new effort was made, in keeping with years of Soviet propaganda, to put the blame on the West.

Western scientists were the first to develop the atomic bomb. A Western power, the United States, was the first to use it in war. The Western alliance, the North Atlantic Treaty Organization, has based its strategy on the assumption it will use nuclear weapons to meet an aggression. This decision was reasserted on Thursday by Field Marshal Viscount Montgomery, Deputy Commander in Chief of the Atlantic forces in Europe.

Thus the Western powers are faced with a moral and strategic dilemma that troubles the consciences of many in Europe.

If the West foregoes use of nuclear weapons it will be hopelessly inferior on land to the Soviet Union in Europe and to Communist China in Asia. Both these powers have enormous armies that the West cannot match. The West is not even trying to match them but, as the recent British defense decision showed, is reducing armies in favor of nuclear weapons.

THE CONTAMINATION PROBLEM

But if the Western powers adhere to their policy of basing their defense on nuclear weapons in the hope of thus preventing war, they will risk playing into the hands of Soviet propaganda and causing the vast uncommitted Middle East and Far East to believe the West is responsible for a nuclear arms race.

At the same time the Western governments probably will face growing criticism from their own peoples, who fear not only a nuclear war but even the periodic nuclear bomb tests that contaminate the atmosphere that people breathe and the plants and animals they eat.

Western leaders have sought to limit armaments, including nuclear armaments, but have been unable to find any system of inspection to insure enforcement of any agreement. But it is urged in Europe that no such system is needed to drop bomb tests, since all such tests by the Soviet Union are immediately detected and announced in Washington.

Much concern was felt throughout the country last July (1956) by the sensational testimony given by Lieut. General James A. Gavin before the senate subcommittee investigating air power. Lieut. General Gavin was asked by Senator James F. Duff, of Pennsylvania, what in his opinion, would be the effect of hydrogen bombs dropped on Russia.

Gavin replied that deaths would run into several hundred millions, and, depending on which way the wind blew, would extend well into the Japanese, and perhaps the Philippine Islands areas or well back up into Western Europe.

That testimony brought forth this reply from Senator Mike Mansfield, of Montana: "The horrible conclusion given by Lieut. General Gavin should make us realize that in this difficult era, with the world's scientists running riot, we are faced with an uneasy peace based on the idea of mutual terror.

"It would be my sincere hope that the administrations of the world will take into the most serious and deliberate consideration the dreadful reality of Lieut. General Gavin's reply and explore ways and means through which the peoples of the world can be protected from terrible devastations over which they have no control."

These reports from scientific sources speak fluently regarding the danger to every human being if radioactivity is used. After reading them do you feel assured that you and your children are safe from it? Are you willing that it shall be tried out, knowing as you do, that the human race would be the guinea pig for the most dangerous experiment ever tried by man, *and his last one,* if he is wrong?

In reading these statements you probably noticed that the small experimental plant at Arco, Idaho, blew up. It could not even be approached for six months afterwards. Accidents happen to a big plant as well as a small one. Does it make you feel comfortable to think that a plant like the Hanford one, or the one to be built near New York, or Detroit, might blow up?

When you think of the world concern and care given to protect people from the two pounds of radium, which was all the world had twenty years ago, does it make you feel comfortable to know that 8,000 tons of much more deadly uranium salts are produced each year from a present stock pile of 600,-000 tons of uranium ore? Also, do you feel that it is quite safe to continue multiplying the effects you have read about for a hundred, two hundred, or three hundred years, which we would have to do if radioactivity is to be the next world fuel? Do you realize that the government is spending $2,000,000,000 a year in atomic experiments, and private industry is spending an equal amount, or more, for a dangerous and costly fuel when it could quite easily have a perfectly safe and inexpensive fuel in abundance for not more than $100,000 and a little knowledge of the true nature of electricity?

Before making your own decision whether or no you would vote in favor of radioactivity, read once more what Dr. Ralph Lapp said about the atom bomb tests distribution of strontium in the atmosphere. He said: "That at the present rate of H. bomb tests, if continued and probably accelerated, by 1962 there would be enough radioactive strontium committed to the stratosphere to produce a 100% maximum permissible amount in every person on the planet." Do read his report with great mental concentration. No other report gives much concern to what he calls "hold up." Again we say that our primary danger is in that stratosphere "hold up," and not in the "fall out." Dr. Lapp estimates that this "hold up" will fall at about 10% a year. If this is true the human race could not survive another ten years of bomb tests such as the United States and Russia are conducting.

We are not in full agreement with Dr. Lapp concerning the 10% fall to earth, for the potential of that "hold up" is lower than the potential of a pressure ten miles up, and it is slowly expanding to still lower potential. The already too heavy "hold up" will blanket the atmosphere layer and eat into it slowly, but irrevocably, and thus thin our life giving and life protecting blue envelope. If Dr. Lapp is right, the human race would already be doomed.

Again we urge everyone to obey the advice given by Dr. Lapp, which we quote from his report as follows: "Education about the biological effects of nuclear radiation is an absolute prerequisite if the human race is to survive."

PART II

PRELUDE

In the heart of every man is the desire for love—for happiness, peace and prosperity. In the hearts of nations of men is the One-World desire for international unity. Universal Brotherhood is the ultimate world-goal. How far are we from that goal? The distance is far ahead for the human race. Its distance is measurable only by the yardstick of LOVE, and love has not yet entered the doorway of human relations, and will not until unity opens the door for it to come in.

Unity means Oneness. Unity balances the mate-hood of all pairs of people and equalizes their transactions in giving and regiving. Civilization has not yet progressed to the point where UNITY, through the GIVING and REGIVING of LOVE, can even be comprehended by more than a few of man's millions.

Man's greatest lesson is to learn how to give. We are still in the barbarian age of taking—of even taking life. The spiritual age is dawning, however. Its doors are slightly ajar, but that is all, for our World-House is erected upon the sands of disunity. The product of disunity is fear. A house divided against itself by disunity and fear must fall. Peace, happiness and prosperity cannot enter a divided house of fear.

If you now ask when peace, happiness and prosperity will come to man, we say to you that it will come when the Light of Love unites all men as ONE, and that shall come to pass only when man ceases to create fear.

CHAPTER IV

The True Nature of This Mind and Motion Universe

I

THE TIME has come in man's unfolding when he should have more knowledge of the nature of this electric universe of motion. He has never truly known it, for he has but sensed its shadow. The real meaning of substance has never been known by man for what it is. He only knows it for what he *senses* it to be.

Actually we live in two universes—the invisible zero universe of CAUSE, and the visible universe of EFFECT. We have sensed the EFFECT and believed in its reality. *We have never yet known the universe of CAUSE.* It is time that we begin to know God's invisible universe which is in absolute control of the visible universe. Man will never solve the riddle of the universe until he fully knows and comprehends the zero universe which he can in no way hear or see. He can know it, however, and in so knowing he can, likewise, know God. He can even prove the fact of the omnipresent God in his laboratory. Nor shall man ever solve the riddle of his own Self—his own Identity, until he knows that he, himself, is as eternal as God is eternal. When man knows that he is Mind and that his body is his Mind-Creation, as the whole universal body is the Mind-Creation of God, he will then know what the consummate mystic of two thousand years ago meant when He said: "I and My Father are ONE."

That is man's first great lesson. For ages he has believed that his body is his Identity, and that his Mind is *in* his body but limited to the boundaries of his body. It has never occurred to him that his Mind is IN—but not OF his body. It has not yet occurred to him that his body is but a "tape recording" of what his Mind is "dictating" into it. He has not yet begun to realize that there is but ONE CREATOR in this universe, nor that he is ONE with that ONE. When man's first great lesson has been learned through the awakening of that divine Cosmic spark, which has for so long been dormant in him, he will then have reached his first stage of God-awareness which all men seek, even though long ages pass before any man even suspects that he is seeking the Light of his divine Self. *When that great lesson is fully learned he will then know that he lives eternally and cannot die—and that his body manifests his thinking for awhile, then rests for an interval before again manifesting his thinking from the point where it rested in his yesterday.*

II

As a prelude to this very brief outline of the principles upon which our universe is founded we will reduce that entire principle to one sentence, from which we will not depart during this entire narrative. We will but expand this one basic principle by describing its processes. Herein follows that one sentence:

The Father-Mother of Creation divides His sexless unity into sex-divided pairs of father and mother bodies, for the purpose of uniting them to create other pairs of father and mother bodies in eternal sequences forever.

This is the whole simple plan of Creation. We will now begin to unfold the seeming complexities of its really very simple processes which so mightily deceive our senses. If our comprehension is strong enough we will be able to continually see a simple thread, which is repeated over and over in every effect of motion throughout all Nature, and throughout every event and experience of life. We will endeavor to enable you to see that simple continuing thread by repeating it many times in many ways.

III

Human progress is always marked by change. As civilization unfolds—or grows—its concepts change to keep pace with that growth. Ages of marked wisdom become archaic to ages of greater wisdom. The last two or three decades of progress have moved the world ages ahead of the last century, and the last two or three years have promised to open new doors into that invisible universe where The Creator awaits man's recognition and awareness of His omnipresence.

Science is discovering God's invisible Magnetic Mind universe through little peepholes into it through a newly dawning Intelligence, but science is not yet aware of what it is discovering. Science calls it anti-matter as yet. That is wonderful—and hopeful. To have discovered and believed that there is a *something* outside of matter, which is seemingly the source of matter emerging from it, is so wonderful that it seems like a thousand years of progress in a day.

Matter is *motion.* The anti-matter, which now engages the serious attention of science, is *stillness.* Some scientists say it is *pure energy.* If matter, *which is but motion,* is energy, and anti-matter, which is *not motion,* is pure energy, what kind of energy is that which is impure? What does it mean? *The time has come when it is imperative that science must divorce motion and energy as one identity, and regard matter as but the product of the Energy Source.*

We wish to demonstrate to you that the anti-matter stillness, which science is discovering, is the Creator's Energy Source. Yes—there is an invisible zero universe. It is the Mind-universe of Magnetic Light. For scientific purposes in explaining the construction of matter, we will name it the *omnipresent universal vacuum. The universal vacuum is the expansion end of the universal piston, and gravity is the compression end.*

As electricity is the creator of focal points, which we call gravity, and because compression is the sole office of electricity, every oscillation of the electric current of Creation is an interchange between the stillness of the universal vacuum of God's Mind-universe of CAUSE, and the electric universe of motion, to produce EFFECT.

Each interchange between the invisible omnipresent universe where motion begins and ends, and the visible transient universe, which multiplies and divides the speed and power of motion, is a cycle. Pulsing cycles constitute the heart beat of this universe, which simulates Mind-Idea through pulsing cycles of two-way motion. Every pulsation of the life principle of multiplying motion creates divided electric male and female bodies, which seek rest in each other from the strains and tensions of their division into pairs of opposites.

The supreme effort of this book is to have you comprehend that Creation is a division of The Eternal One into countless twos of sex-conditioned opposites, which eternally seek to void their division by uniting as One.

Creation is a Mind-imagined journey of electrically divided male and female opposites in increasingly greater speed and power in the direction of each other, where they unite in rest to re-divide, and return with ever increasing speed and decreasing power, to the zero of rest in the Magnetic Light of Mind from which they sprang into action. *Creation is an illusion which stimulates substance by multiplying the speed of centripetal motion, and loses its appearance of substance by multiplying its centrifugal speed.* If we prove this new state of facts to be in accord with Nature's plan, it will obviously be necessary to have very different concepts of the Nature of the universe than those concepts which are now held by man.

IV

We herein print two diagrams to symbolize both ends of the compression-expansion pump, which this universe is. We cannot move our little finger without gaining the power to do so from the vacuum Source of that power. We do not know that fact yet. We do not yet know that we live, and breathe, and express our multiple desires only by constantly taking power for each cycle from God and giving it back to Him at each cycle's end. Man has not yet become aware of the fact that he is but the motion within the divine Cosmic vacuum tube which sparks its little light from out of its dark, and is, himself, that dark when his light goes out. *When man is fully aware of the fact that he eternally lives in God's invisible Magnetic Light as One with it, and merely manifests life by action in the electric universe, he will then know that when action ceases it merely ceases without affecting him, even as sound ceases without affecting him who made the sound.*

One of these drawings symbolizes the Energy end of the Cosmic Piston from which gravity emerges in moving pairs to simulate energy by motion. The other drawing symbolizes the effects of motion at the compressed end of the Cosmic Piston.

V

It is self-evident that all motion springs from rest and returns to rest for eternal repetition in sequences which we call electric frequencies. It is also self-evident that all motion springs from out of the dark in divided pairs and multiplies their gaseous beginnings into dense bodies of incandescent suns, then disappears into the dark through the gases of themselves.

WHERE and WHAT is this invisible dark from which the light of motion appears, and into which it disappears? Is not this the supreme question of all the ages? Can there be any other answer to that question than the fact that Creation emerges from its Creator?

The greatest thinkers in science have repeatedly said that matter emerges from *space* and is swallowed up by *space* in some unknown and mysterious manner. The word *space* is rather a casual word to use in place of the Creator. Likewise, it is a misleading word, for space is not an expanse of something outside of matter. It is as omnipresent within matter as it is without, and it is in control of matter from within as well as from without.

Future generations of enlightened men will cease thinking of this electric universe as being matter and substance. *They will know it for what it is, which is motion only.* When the idea of substance has passed out of our thinking, and simulation of substance by motion will have taken its place, mankind will then comprehend his Mind-unity with the Universal Mind. And he will know his body for what it is—merely an instrument for creating form-images of Mind-thinking.

The Creator is the Supreme principle of Unity in His Oneness

CAUSE

FIG. 6 symbolizes the uncreated universe, the zero vacuum universe of rest, from which the objective electric universe of motion emerges to manifest eternal life by simulating it in cycles of motion. This is God's omnipresent universe of Idea, and Desire for the expression of Idea, by moving forms which symbolize Idea. This is the universe of The Imaginer Who builds forms in the image of His imagining for giving and regiving in cyclic intervals, which unknowing man calls life and death.

This is the omniscient universe of eternal MIND-QUALITIES from which transient electric QUALITIES emerge to simulate the qualities of Mind—Magnetic Light—Knowledge—Idea—Energy—Life—Soul—Love—Truth—Beauty—Rhythm—Balance—Law—Silence—Rest and Stillness.

This is the abiding place of eternal MAN who is ONE with GOD. It is God's kingdom of heaven within man, which the Nazarene bade men to seek.

of Being. The supreme urge back of His divided pairs is to seek unity in each pair. This is as true of the elements of matter, as it is true of living organic beings. Since all matter is divided into mate-pairs, is it not time that the Father-Motherhood principle of Oneness in the Creator, be recognized as the desire to attain that father-motherhood oneness by all of the divided mate-pairs of all Creation? Every particle of inorganic and organic matter in the universe is perpetually seeking unity with its balanced mate. There is no exception to this fact in all Creation. Sodium seeks balanced unity in chlorine. It does not seek it in aluminum, phosphorous or silicon. Nor do any of

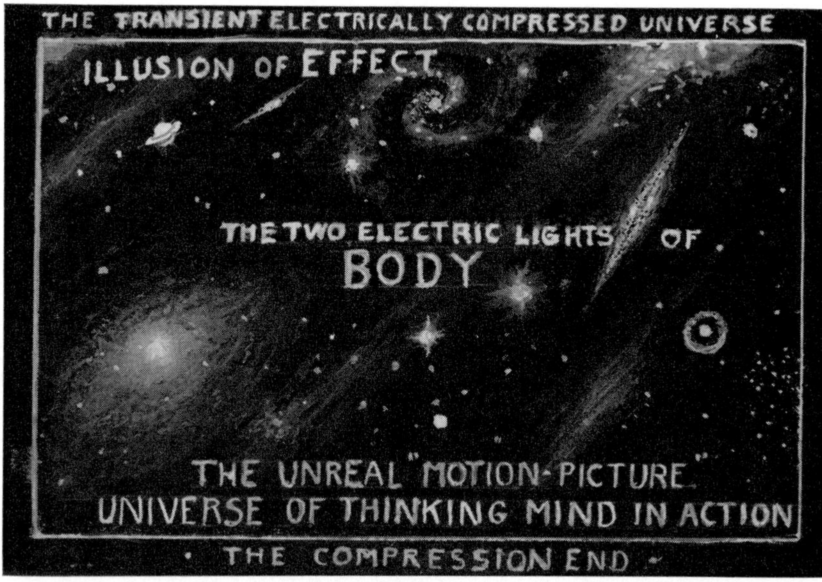

EFFECT

FIG. 7 symbolizes the transient, created quantities of divided pairs of units of electric motion which emerge from the vacuum zero universe to manifest its qualities for timed intervals, and are then dissolved into its equilibrium for re-emergence. These seeming quantities are Electricity—Compression—Form—Light-spectrum—Expansion—Sex—Sound—Heat and Motion.

These multiple quantities seem to exist. They appear and disappear to again reappear, to simulate the qualities of life, energy and idea which they seem to be, but are not. This is the "motion-picture" universe of simulation and illusion which dramatizes the idea of REALITY by electrically projecting the thought-imaged-forms of Mind—Idea upon the imagined-screen of space, which Nature is.

There is no life, energy, knowledge, idea, truth, intelligence or any other quality of Mind or thought, in the motion which man calls matter and substance, and mistakes for reality. It is time that man realizes that there is nothing, whatsoever, in this visible universe but motion, and that REALITY exists only in the vacuum from which motion emerges as heat to simulate the Light of Reality.

the elements on the red, alkaline side of the spectrum, seek union with each other, nor do those on the blue, acid side seek union with each other.

If this search for balanced mates is universal in all bodies of matter, and if the division of an equilibrium into polarized pairs is the effect which electricity produces, why do we not recognize the sex principle as the basis of all electrically divided, and pressure conditioned matter, and discard the old text-book idea of two kinds of electricity, *one which repels and the other which attracts,* as being the basis of matter and motion? See Fig. 8.

The Coulomb law would be more understandable under this concept, for, naturally, males do not seek union with males, and two red, or two blue sides of the spectrum do not make a complete spectrum. Such an inconsistent concept should be no longer applicable in a progressive world of slowly unfolding intelligence. The Coulomb law of attraction of opposites and repulsion of likes, has no relation, whatsoever, to Natural law. Nature does not attract, nor does it repel. Pairs of opposites in Nature are projected centripetally toward each other by the light mirrors of wave-field projectors of light images, which must await further explanation in the very brief description of the geometry of space that we can include in this book. This basic new science is unknown as yet but we are preparing to leave full records of it for posterity if—and when—posterity shall be ready for it. Suffice it for here to say that all mates which are approaching each other for union are electrically forced to approach each other from the outside. They are not "attracted" together from the inside. They cannot help being forced. They are being wound up together electrically in spiral vortices. They must inevitably collide and become one even as a threaded screw is forced to interpenetrate an oppositely threaded nut. This simile is a good one in other ways. It exemplifies the red and blue of the spectrum and the red and blue of sex division. The red interpenetrates the blue in all elements up to carbon. Beyond carbon the blue interpenetrates the red. This is also an unknown basic characteristic of the creative process, for no text book of science even mentions it.

The Coulomb law has been fundamental for decades. Those who still desire to cling to it must deny the Einstein Equasion of 1905, for both of these contradictory laws cannot be true. *One or the other must be wrong.*

VI

All motion in all this universe is based upon the fact that the sexless condition of stillness has been divided into two sexed

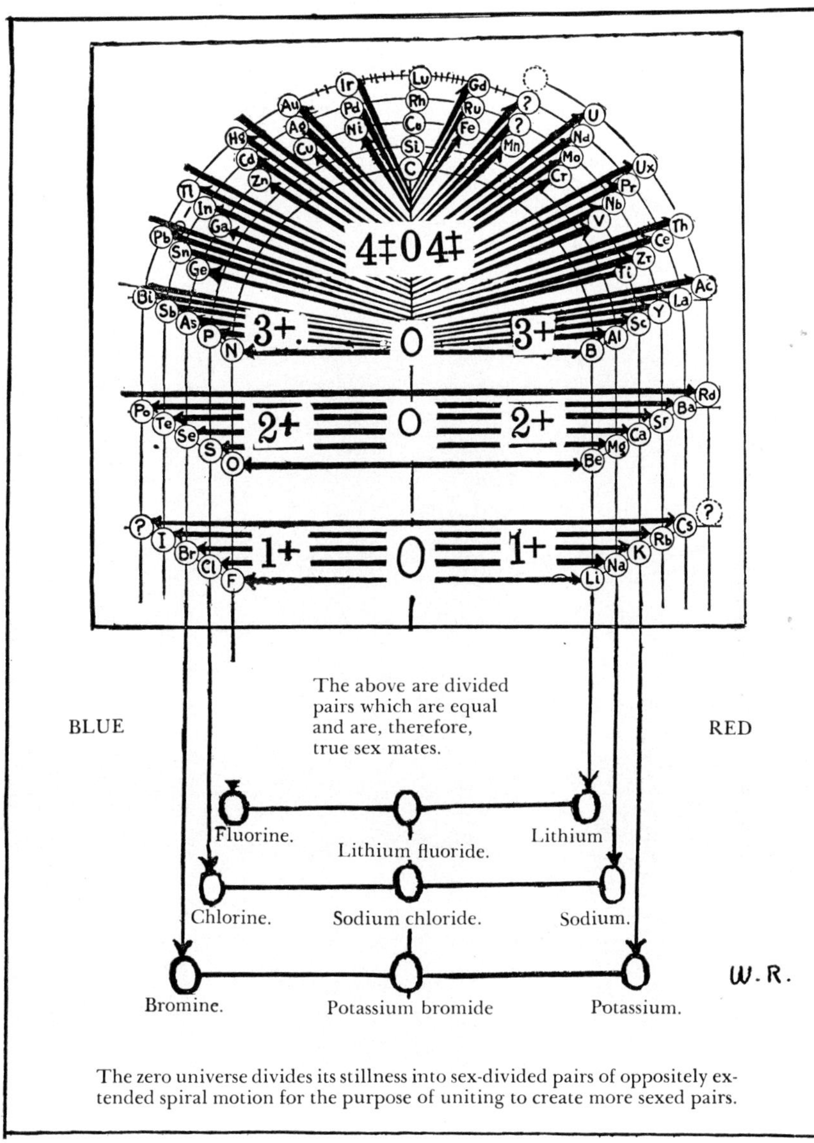

The zero universe divides its stillness into sex-divided pairs of oppositely extended spiral motion for the purpose of uniting to create more sexed pairs.

FIG. 8. Nature knows no other process than sex division and sex unity.

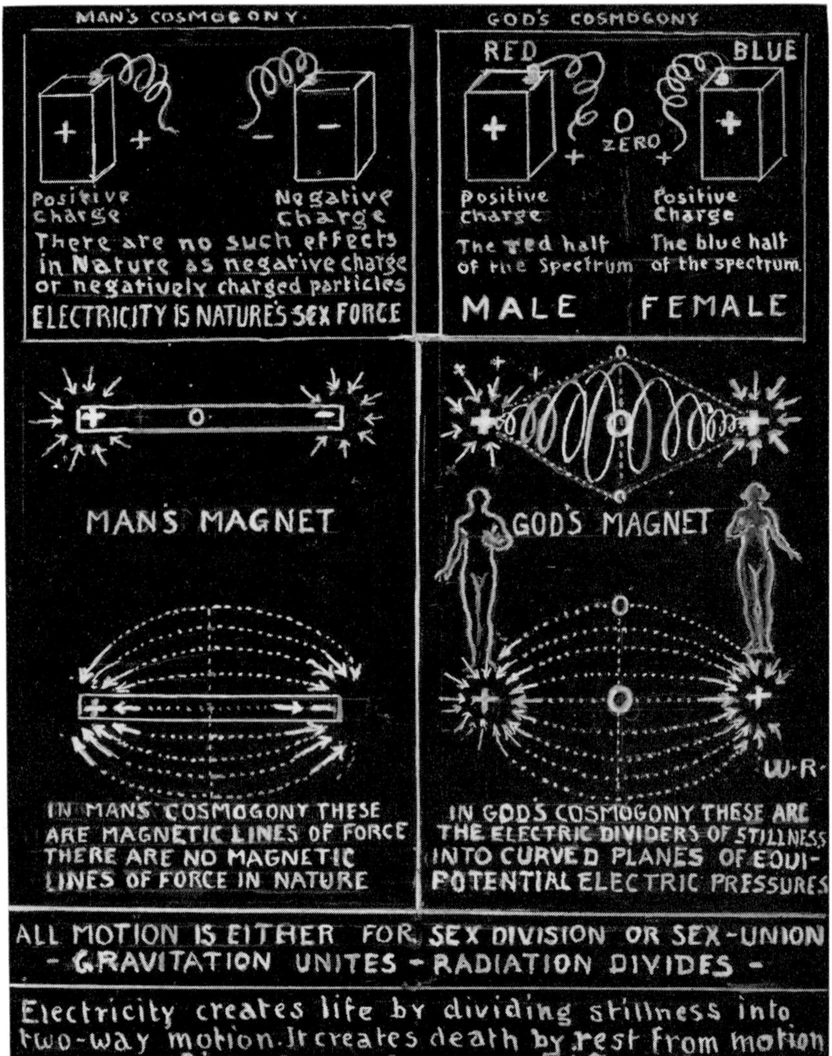
MAN'S COSMOGONY
GOD'S COSMOGONY
Positive charge
Negative charge
There are no such effects in Nature as negative charge or negatively charged particles
ELECTRICITY IS NATURE'S SEX FORCE
RED
BLUE
ZERO
Positive charge
Positive Charge
The red half of the Spectrum
The blue half of the spectrum
MALE
FEMALE
MAN'S MAGNET
GOD'S MAGNET
W.R.
IN MANS COSMOGONY THESE ARE MAGNETIC LINES OF FORCE THERE ARE NO MAGNETIC LINES OF FORCE IN NATURE
IN GODS COSMOGONY THESE ARE THE ELECTRIC DIVIDERS OF STILLNESS INTO CURVED PLANES OF EQUI-POTENTIAL ELECTRIC PRESSURES
ALL MOTION IS EITHER FOR SEX DIVISION OR SEX-UNION
- GRAVITATION UNITES - RADIATION DIVIDES -
Electricity creates life by dividing stillness into two-way motion. It creates death by rest from motion

Fig. 9 · Fig. 10 ·

conditions. The division begins its tension at the cathode and multiplies that tension as the division extends from the cathode toward the anode. Electricity thus creates tensions and strains, which we call electric potential. Electric potential is the measure of compression at any one point in the universe. Gravity is evidenced wherever compression from without is maximum. *Gravity is a focal point from which matter desires to explode outwardly. Gravity does not pull inwardly from within as the deceptive illusion of Nature would have you believe.* Neither is it the attractive force which Newton's senses were deceived into believing, for a center of gravity is a point of maximum electric potential. Gravity never changes. It is never more or less. It is maximum everywhere. It is electric potential which changes by multiplying and dividing the measure of power it is able to express by drawing it from its gravity center of control. It is like the power of a man who can draw but little from the great omnipotence which centers him, as compared with another man whose knowledge enables him to draw more of it. *Electric potential is the tension caused by the electric division of the ONE universal condition of rest into two unbalanced conditions of motion.*

Polarity and sex are one. The two poles which unite are male and female opposites in all matter. The idea that sex is solely a characteristic of organic bodies is not a true concept. Sodium and chlorine are as male and female as man and woman are male and female. Every particle of matter in the universe is either male or female, and these pairs are electrically projected together in the ratios of force which Newton gave to the acceleration of gravity, and upon a multiplication of speed and potential principle, which Einstein gave in his Equation of 1905.

If, therefore, all motion is solely for the purpose of either seeking sex union or sex division, why should we not immediately remodel our thinking to conform with Nature's processes and now think of our divided, double pressured universe as an electrically sex conditioned one?

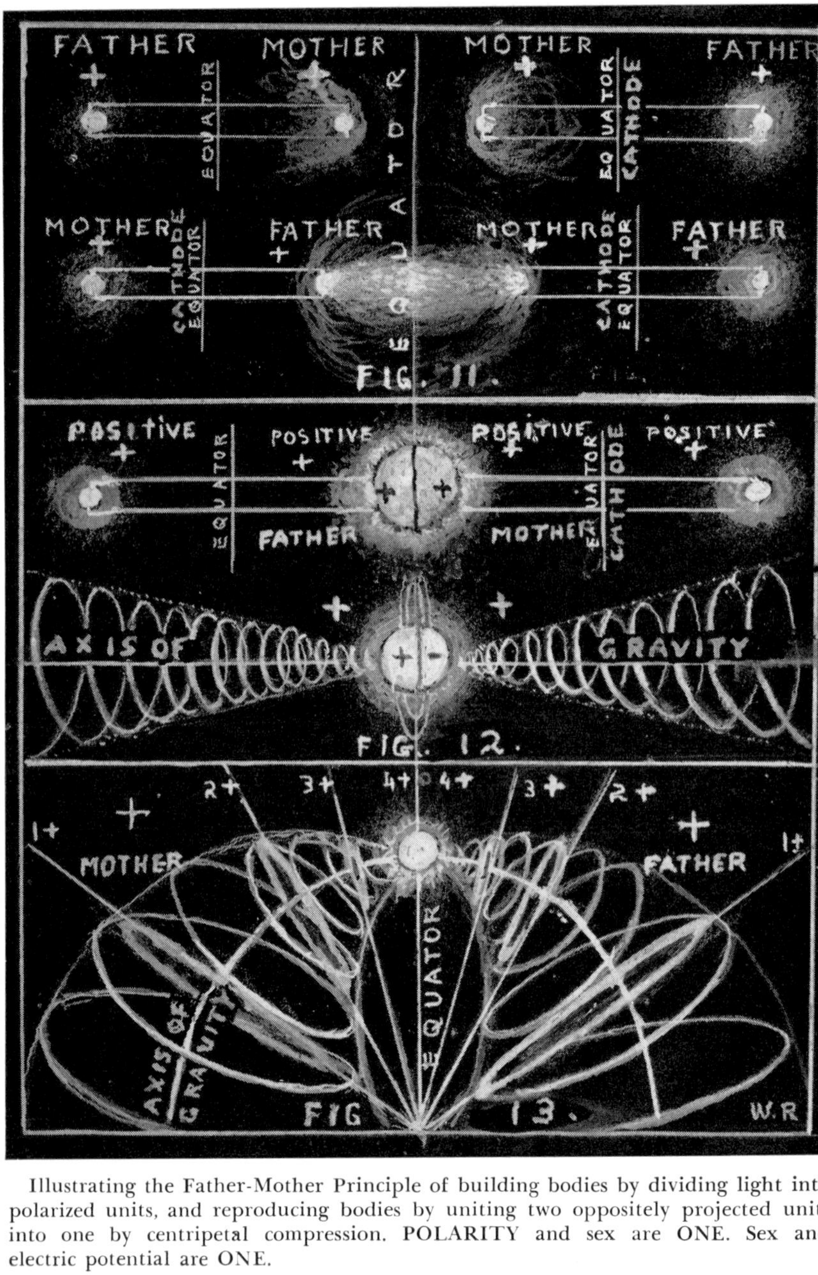

Illustrating the Father-Mother Principle of building bodies by dividing light into polarized units, and reproducing bodies by uniting two oppositely projected units into one by centripetal compression. POLARITY and sex are ONE. Sex and electric potential are ONE.

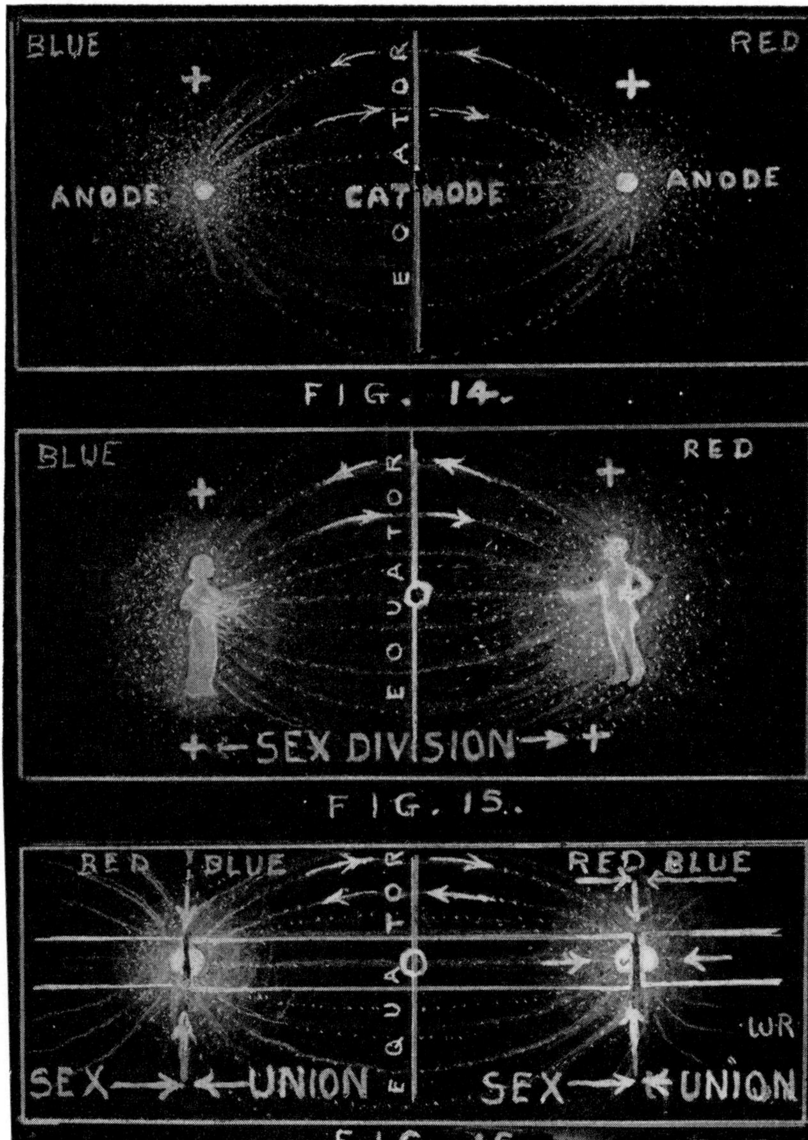

The cathode in Nature is the still point of Magnetic White Light of Mind from which motion begins to compress in two opposite directions to create sex-divided bodies. Gravity begins when compression begins. Polarization and compression are ONE.

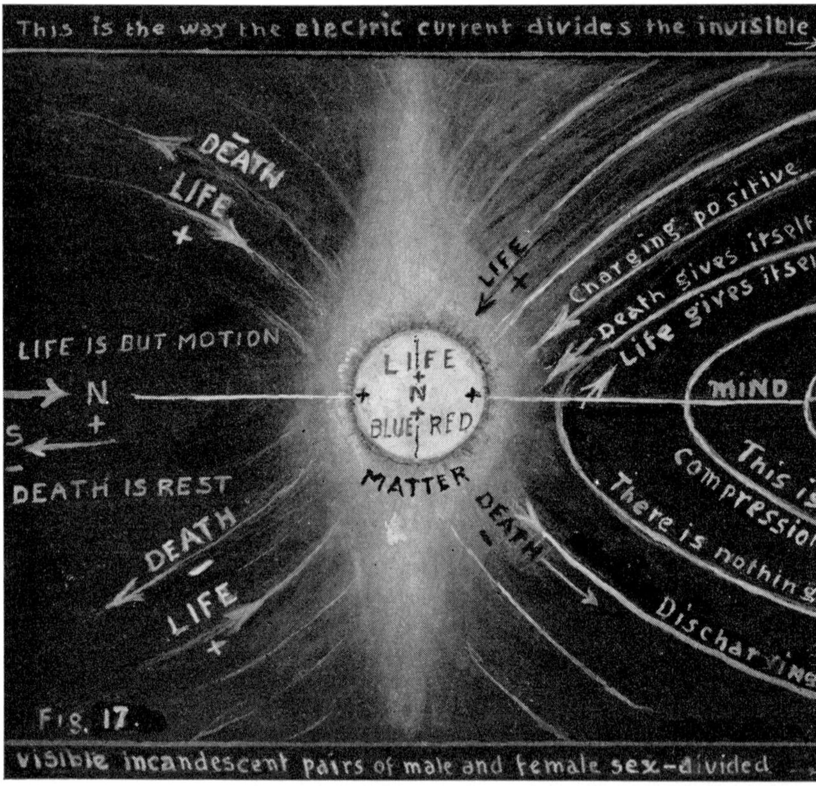

This drawing shows the method and process of interchange between the zero universe of Mind and the multiple universe of motion. Study all other drawings and relate them to these two. The more you comprehend them the more you will also comprehend that life is but fast motion and that death is a cessation of motion.

VII

The division of sexless Oneness into pairs could not be a part of Nature's process without also dividing the Oneness of Light. God's Magnetic Light is white because it is still. It has no tensions or strains in it, for it is not divided. It is invisible to the senses because it has no motion in it. The moment motion begins the Light of the Creator's energy is divided into pairs which multiply their red and blue intensities in the ratio of their extension from their cathode beginnings, just as electric potential multiplies its intensity for the same reason. If, there-

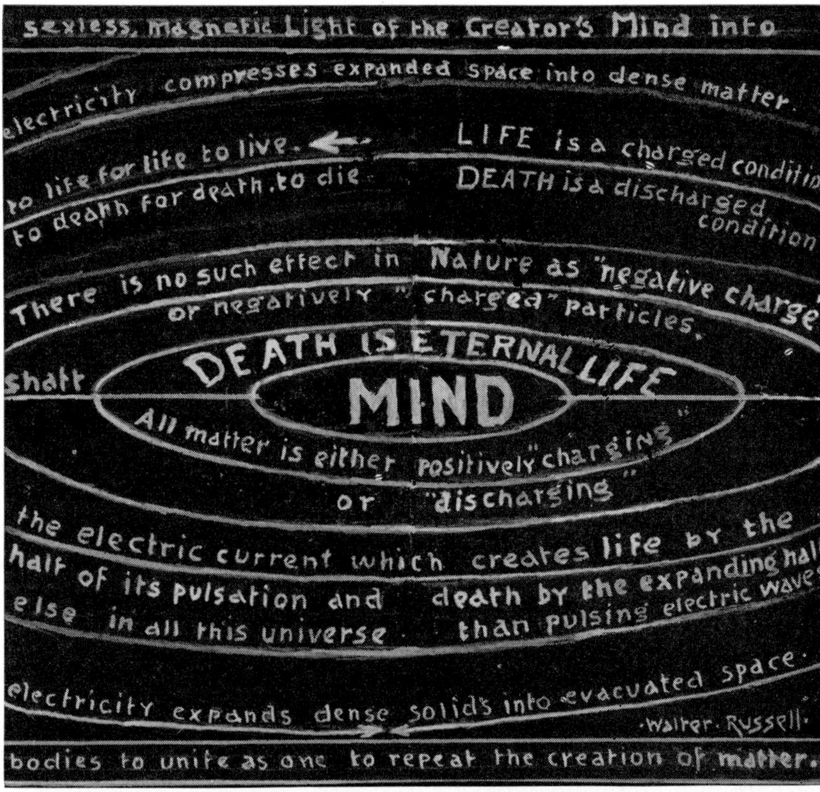

This shows how the heartbeat piston operates between the vacuity of the Creator's Magnetic Light universe of Mind-energy and the potential of the electrically pressured universe to create cycles of two-way motion. The more you can comprehend the secret of God's Mind-control over His Creation, which is His universal body, the more you will be able to control your body, for you will then know that God's Mind and yours are ONE.

fore, the Creator electrically divides His One pressureless, motionless Light into two oppositely pressured lights of motion, is it not imperative that we associate these two similar effects which rise from the same cause, and judge them as ONE. Should we not recognize the fact that these two colliding light pressures wind up together, as a bolt threads into a nut, or more properly as one spiral spring intertwines into another? Is it not logical, therefore, to realize that the interpenetrating half of the spectrum is red, and the outer half is blue? That is the way we

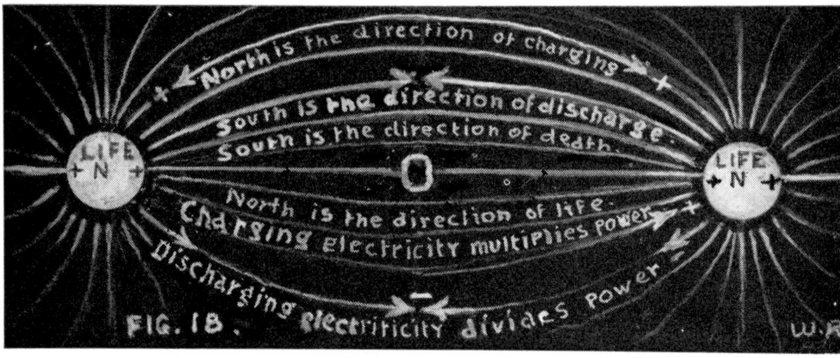

A study of Fig. 18 together with the bar magnet diagrams and other polarity diagrams shown in Figs. 11 to 16 will prepare you for a better understanding of the nature of electricity and the electric current.

find it in Nature. The inner core is the red of its heat and the outer layers are the blue of space into which the red core projects its heat to expand it. The outer layers then cool it and thus create imprisoned bodies whose sole desire is to escape imprisonment. Beyond carbon the inner core reverses to blue.

That is the way it is all through Nature. Compressed motion heats bodies which endeavor to escape. Expanded cold freezes bodies to keep them from escaping. That is the principle of body-forming. That is the principle of growth from a seed to a tree. The seed generates heat from the moisture of the sun-warmed ground. The seed explodes in its endeavor to escape from its compression prison. The explosion creates heat which is cooled and that which unfolds is refolded. Cell by cell is thus added by the interchange of motion between red light and blue, and between heat and cold as the father interpenetrates the mother womb of Nature to generate bodies, and the bodies escape from their wombs to return to their source.

As we thus unfold Nature's processes and her way of working you will gradually comprehend that there is no place in God's Cosmogony for such a concept as negative electricity, or negatively "charged" bodies.

We live in a universe in which all creating, charging, polarizing, compressing and heating bodies manifest LIFE, and all discharging, depolarizing, expanding and cooling bodies manifest

DEATH. One cannot say that a dying body is negatively charged, for a charging body is positive. It is living. Your car battery can be charging from its generator, and simultaneously discharging, from its use, but if you had no generator in your car the battery would discharge its charge until electricity ceased in it. While it is dying you would not even then say that it is negatively charged, for it would still be positively charged until it could no longer move. We do not say "negative sound" or "silent sound" when it is ceasing, or has ceased. Nor do we say "negative warmth," or "cold warmth," when a hot iron is cooling, or has cooled. Nor do we say "negative life," when a man is dying, or is dead.

Electricity compresses and ceases to compress, but we do not then say that it is "negatively compressed." It is for that reason we put a positive sign wherever an electric condition is existent, but there is no need for any sign unless there could be two which indicate sex, and two which indicate charge and discharge. *When motion ceases electricity ceases. It is not then negative electricity—as death is not negative life.* In all of these charging and discharging effects the tensions of the color spectrum parallel the tensions of electric potential. That is why they should be considered together.

If this change in our thinking is true to Nature, there can be only one kind of electricity—positive electricity, meaning the power to charge or compress. It necessarily and logically follows that a negative electricity would be impossible in an electrically polarized universe in which both divided mates are equal. It also would be unnecessary if the other end of the compression pump, which this universe is, is the universal vacuum. Shall we call this *new thinking,* or is it but *changed thinking?* Shall we concede that early observers who arrived at such conclusions in the infancy of scientific research, were right, without the possibility of being mistaken? Or can we not consider them as torch bearers, rather than anchors?

It is in this sense that we present quite a different picture of the universe than the old one, which early observers sincerely and conscientiously gave to us as the best they knew in those early days. We cannot present the new concept of the Nature of the universe in the terminology which sprang from old concepts, however, so we beg of our readers that you realize the necessity of so doing is not critical of past contributions to world progress, but are only adding to it by changing old concepts to fit new illumining, which time and research have made possible. We feel that this explanation is necessary because some of these new changes of concept are so radically different from traditional training and thinking.

We repeat, however, that the main purpose of this book is to give this new knowledge in the interest of world-survival only at this time, and is not given as a text book of new scientific thinking for educational purposes. That will come later, and in other ways, such as we are gradually giving to our students. For this reason these writings are brief, but brief as they are, they will throw more light upon the new age of transmutation, which would surely follow a change from the old concepts of matter and motion to the new concepts briefly outlined herein. Transmutation needs but the comprehension of a few men of vision, men whose minds are not fixedly closed to make it a reality.

Transmutation is the simplest of all the principles of Nature—ten thousand times more simple than what is now being dangerously, and expensively done. The first step toward it is to discover that matter and motion is tonal—and thoroughly controllable. The elements are not substances. They are but states of motion, which can be controlled by knowledge and electric power. The present concept, which makes it necessary to knock out an electron from one element to transmute it to another, is contrary to Nature's working. No one would think of trying to knock out one part of a sound to make it into another sound, or to take an arm from a man to transmute him into a horse. That is what science is trying to do, however, by knocking an electron from mercury to produce gold. There are no electrons in the elements to knock out, in any case, as we shall see.

Nature's patterns, species and sexes can be modified but they are the same in kind. Transmutation means modification through knowledge of how to control that modification, which is as simple as the retuning of a harp string by increasing its vibration frequencies. Nature divides, multiplies and combines by the use of electricity. Man has all the electric power he needs at his command. Every state of motion, and any combination of those states, can be made use of by man if he but knows how Nature does it. Every element can be retuned or divided into pairs, and the pairs retuned. Every combination, such as the atmosphere, can be divided and its pairs used separately or recombined as one wishes. Hydrogen, oxygen or nitrogen, in combination, can be taken apart with ease and used separately, or recombined, or modified at will. Likewise, salt water can very simply be taken apart and recombined minus its salt, at will.

When Nature divides she always creates male and female mate pairs. She then multiplies those mate pairs separately, or in union. She multiplies sodium into potassium, then into calcium. She multiplies fluorine into chlorine, then into bromine and again into iodine. Man could do this same thing with any tonal element and even split them into semi-tones. Man could also divide such elements as carbon and produce, from carbon alone, five pairs of rustless, stainless metals of high melting points, high malleability and conductivity. The elements are but the alphabet of man to write what he chooses with its letters. They are but the tonal notes upon the nine octave keyboard of the

Cosmic instrument with which he has but begun to compose the symphonies of his desiring and his imagining.

Nature can create cyclones if it becomes conditioned for cyclones, but man can control those conditions and divide them at his will. Likewise, man can create rains wherever he wishes and in any amounts. The world's vast deserts can be forested with date palms and carpeted with a verdure which would end dust storms forever, and add materially to the world food supply. To enable man to do this he needs only the knowledge of space-geometry, mathematics, the wave of gravity control, and the nature of electric current.

The great power to transmute through Mind-*control* by Mind-*knowing,* is man's when he finally knows *CAUSE* instead of being limited to the *effects* of *cause.* When that day comes man will no longer need to use the limited supply of the earth's fossil fuels, nor need he labor to procure them. The age of transmutation can be but one year away if man chooses to open his mind to new concepts, or, otherwise, it may be three thousand years away. CAUSE can never be known by the study of its EFFECT in motion. Cause is in knowing Mind, not in sensed-body. Cause lies within the invisible universe which does not respond to the senses, and not in the senses *which can but sense motion and can never know.*

The senses forever look out through the convexity or concavity of pressure lenses and mirrors of a curved universe of two-way motion. Distorted images and an upside-down universe is recorded upon the senses. The inner vision is not electric. It is Magnetic. The cube mirrors of space are of zero curvature and do not distort. The Mind-conception transcends the sense-objective, and is not deceived by illusion, for Mind is the creator of illusion. *Mind knows and projects its Light. Senses can but reflect. They cannot know.*

And so it is that man peers into microscopes, and telescopes. He builds great laboratories for research into effects of things which move. He gathers much information about their movements and develops great skills in controlling the movements of moving things. He then reasons and assembles, *but reasoning is not knowing—and assembling is not creating.* But what does man KNOW? Informing the senses of effects of motion does not awaken knowledge in Mind. On the contrary all effects of motion are optical illusions, which deceive the senses and cause men to form conclusions which are not true to Nature.

It is now time that man must realize that the key to Nature's secrets can never be found in the visible universe. It can only be found in the invisible universe which creates and controls the visible. But we cannot know the invisible universe because we cannot see it," man says. As the spiritual nature of man unfolds he becomes aware of the fact that he can *see* one half of the universe and never *know* it, and the other half he can know but cannot *see*.

We believe, also, that man can never find true happiness until he has finally solved the full meaning of what life and death really are, instead of the dread meaning which he has given to a belief in death, which has no validity, whatsoever, in Nature. For these reasons we shall briefly define God as we have pictured His Omnipresence and purpose in Lesson Number 13 of our "SCIENCE OF MAN HOME STUDY COURSE." This basic foundation of Creation will aid in comprehending the true nature of electricity and gravitation.

CHAPTER V

Prelude—the Transformation of Man

When man shall at last know God he shall then be transformed.

Dynamic knowledge of the Identity and Presence of God within the Soul of Man—and throughout all Nature is a necessary stage for the human race to attain before it is possible for him to build an enduring civilization of unity and brotherhood.

The only reason man has never known God is because he is still in his intellectual infancy and his ideals and practices are still pagan and barbarous.

To know God man must know what Light is, and energy, Magnetism, electricity, gravitation and God's processes in the building of bodies which manifest His Presence and His purposes. And he must know the mysteries of life, death, and growth, and of the seed into which even the giant oak disappears but still retains its identity. And he must know how man retains his identity when he also disappears.

No man has yet known these things. The concepts of modern sages regarding them are totally unlike God's processes. The conclusions of man regarding them are the conclusions of his *senses.* His *Mind* has never yet pierced their illusions. He still senses EFFECTS without knowing their CAUSE.

These things you shall now know—and when you comprehend them within your Inner Self you will then know God.

We shall now devote this entire chapter to the complete understanding of the Identity and Presence of God. It has always been presumed that God is, and forever will be incomprehensible and unprovable, because He is invisible, and that which will not respond to human senses must be beyond human knowing.

That is not true. The invisible can be KNOWN and COMPREHENDED even though it cannot be seen. *One cannot see gravity, but one can know and comprehend it, for one must obey it or perish.* Polarity, truth, silence and love cannot be seen but these invisible realities constitute the great powers which rule the universe, and they too can be KNOWN. Climaxing all these is your Mind. You cannot see your Mind, nor ours, but you know that your Mind is the basis of you, and that it controls your visible body. Likewise, you cannot see God's Mind yet you know that this universe is His creation and that He rules it with a power which you also cannot see, but you must obey it together with all creating things.

Primate man demanded a God who could be seen. He worshipped the sun, then made idols. Our present civilization is still pagan. It has never passed beyond that stage. The great mass of mankind still conceives a huge objective manshaped God with human emotions of wrath and vengeance.

It is time that the human race should know God as He IS. A happy, progressive and enduring race is impossible until that new dawn of Conscious awareness of His ever-present Being is intimately close to us always. How far away from that day are we? Out of the world's pagan thousands of millions there are but a few who have begun to comprehend that the path to the Light of the inner kingdom is illumined with Love, and that Love is inevitable and irrevocable LAW which no one can violate and survive. Our present almost primitive civilization is barbarian, in the respect that it is attempting to survive, endure, be happy and prosperous by violating God's Law of Love. Man has not yet learned that he cannot violate or break God's Law, but that the Law can break him. When man thinks he is breaking the law by hurting his brother, he is but fulfilling the law of balance which regives to him the hurt which he has intended to inflict upon another.

You shall now begin to know that great mystery which has for ages hidden its face from man. That knowledge of the Identity of God will so increase your spiritual unfolding that your inner-sensory perception will lead you far into the path which leads to the "Brahmic Bliss," which Buddha bade men seek, or "The kingdom of heaven," which Jesus told man to seek. *It is not possible for this age of man to comprehend God unless false conceptions of electricity, magnetism, gravitation, energy and the construction of the atom, as now conceived, are eliminated and replaced with Nature's ways and processes.*

All of man's fundamental conceptions have been the result of forming conclusions which have been based upon the outer-vision of the senses, and not upon the inner-vision of the Mind. The senses see illusion and are mightily deceived. *Mind-vision does not deceive for CAUSE begins there, and EFFECT is but its product.*

The truth of all fundamentals of Nature are just the reverse of the conclusions of science, just as the reflections in a mirror are the reverse of their cause. These conclusions began with someone rubbing amber and glass with wool and silk, and progressed through Newton and other very much deceived observers up to the fantasy of Neils Bohr's impossible atom, which has no resemblance to Nature, whatsoever. The gravity concept at least resembles Nature in reverse, but the Rutherford-Bohr atom has not even that virtue.

When we explain the true nature of electricity we will clearly demonstrate that the familiar model of the atom, which shows electrons moving in orbits of many intersecting planes around one centering nucleus, is an utter impossibility in Nature. It defies every principle of the electric current and the wave, and should, therefore, be relegated to pure invention. It is difficult to describe the shocking effect such a concept has upon an Illuminate who can "see" into the atomic or stellar systems without microscope or telescope, while the outer-vision cannot even discern what holds matter together with twenty million dollar cyclotrons. When you know Nature's working principle you will comprehend what a shock it is to know that it is possible for an enlightened age to believe that electrons in certain numbers revolve around inert gases. When you know what the office of inert gases is in Nature you will be even more shocked.

To exemplify our meaning let us remind you of the familiar belief that magnetism is a force, separate and apart from electricity, which has the power to pick up nails on a bar magnet and tons of iron on a giant magnet. Scientific terminology is redundant with references to such effects as magnetic lines of force, the earth's magnetic field, and electro-magnetism, when every effect attributed to magnetism is solely electric. Furthermore, there is no such separate force as magnetism which performs the work of Creation. *That which man thinks of as a magnetic force is spiritual Light of Mind and not a physical working force of Creation.* Likewise, we hear constant references to negative electricity, negative charge, and negatively charged particles, which are as impossible in Nature as silent sound is impossible.

Perhaps the most fundamental of misconceptions is the Coulomb electric law which says that opposites attract, and that gravitation also is a force which pulls inward from within, and that it attracts other bodies, when, in fact, both of these beliefs are just the opposite from the facts of Nature upon which they were misconceived. Coupled with this unnatural conception is the equally unnatural one to the effect that the universe WAS CREATED about two billion years ago, and is now radiating its energy away, instead of it BEING CREATED eternally. With that concept there is no room for an "uphill flow of energy." There is but a "downhill flow." Knowledge of the nature of electricity would quickly dispel that idea that the universe itself is on its way to death. *There are two opposed actions to* every electric pulsation. One of them is GENEROACTIVE, which multiplies compression. That is Nature's "uphill flow," which charges. The other is RADIOACTIVE, which multiplies expansion, and that is its "downhill flow," which discharges. For this reason it is time that we begin to know the true nature of electricity and Magnetism, rather than theorize from what our senses tell us.

Figure 9 illustrates the modern conception of electricity which is not in conformity with Nature.

When the true nature of electricity is comprehended it will then be possible to comprehend why the Rutherford-Bhor atom concept is utterly unlike Nature. You will gradually understand, that all Nature is based upon the love principle which is expressed by *giving and regiving*. Nature never TAKES. The present concept of gravity is based upon *TAKING*, for it supposedly pulls inward from within itself. Nature does not work that way. Nature does not even "absorb from within," nor are there inward explosions in Nature, nor is there such a force as attraction or contraction. This seems to be an amazing statement but it is a true one, nevertheless, as we shall see when we go back to *cause* and are no longer deceived by the illusions of *effect*.

Man's concept of gravity as being an attractive force which pulls inward from within, is diametrically opposed to the facts of Nature. Gravity is the controlling center of a compressive electric force which is exerted from the *outside* of matter instead of *within* it. As you gradually understand the electric wave, and its enclosing cubic wave-field, and the balance principle upon which all motion is based, you will have different concepts in relation to attraction, contraction, absorption, balance and the universal equilibrium. When you fully comprehend what the invisible Light is, and the two divided lights are, you will also have a different concept of matter.

With this necessary prelude finished we will now proceed to build true concepts as a basis for your comprehension of God and His process for creating the pairs of units of divided light in motion, which are known as matter. So fully has science been deceived by the illusions of motion that it has built up a great field of its own in the scientific world which is commonly known as nuclear physics. As its name implies this concept is based upon the belief that the atom is held together from its inside by a nucleus, and that nucleus is composed of certain particles which act upon each other in some mysterious and unexplained way to hold themselves together and cause oppositely charged particles, known as electrons, to revolve around them. These electrons supposedly revolve in shell-like layers around the nucleus. By adding one more electron to each element it becomes the next element in the series.

There are no particles, or groups of particles, which hold the atom together as nuclei. Gravity does not work that way. All creating matter is centered by holes of space except one element in each octave. *The Energy of Creation centers each hole. That centering, invisible, omnipotent Energy is God's Mind and your Mind. Naturally you cannot see it, but you can KNOW it, for it is your Identity, and your Intelligence. It is the Source of your creations as it is the Source of all Creation.* That is what Jesus meant when He told man that the kingdom of heaven is within. That, also, is what He meant when he said: "My Father and I are ONE."

The time has come in human history when all men must know exactly what these teachings of the great Illuminate really meant. The time has come when man's spiritual unfolding shall awaken the Light of genius, Cosmic Consciousness and Christ Consciousness in all men. Man will never know the happiness and peace of One World of Universal Brotherhood until that day shall come.

The impossibility of a centering nucleus in the atom is because of the fact that *the atom is not pulled together from the inside, it is compressed together from the outside. This is one more example of the deceptions which motion practices upon those who look upon reflected effects with their outer-senses instead of upon the cause of those effects with the inner-Mind.*

As these lessons proceed we will take you right within the atom itself so you can see every one of them as clearly as we see them. A ten-year-old boy could know the atom as familiarly as he knows his alphabet if he but be made aware of the true nature of electricity, and its utterly simple basic working principle, as so clearly demonstrated in the unknown wave in which all the secrets of the universe are locked to the outer-vision of man.

We Define God

I

God is Light. And God is Love. And God is inexorable Law.

God is the invisible, motionless, sexless, undivided and unconditioned White Magnetic Light of Omnipotent, Omniscient and Omnispresent Mind.

In all this universe of countless many things there is but One Light—which is Mind of God—and the two extended lights of His thinking, which His imagining created to manifest His Being in seeming action. There is naught else in all Nature than God's knowing Mind at rest and the motion of Mind-thinking.

God's knowing is Magnetic. God's thinking is electric.

God has but one IDEA, one DESIRE, one PURPOSE, one ACTION and one LAW.

His one *idea* is His unity in Father-Mother Love. His one *desire* is to think action, and rest from action, in sequential intervals. His one *purpose* is to think His knowing of what He IS, and what He KNOWS, into His own image. His one *action* is to give. His one *law* is that *what* He gives must be balanced by equal regiving.

His Law of Balance is the law of Love upon which His universe is founded, for God is Love—and the universe must reflect His image.

God's Magnetic Light universe is at rest. It is balanced.

God's electric universe must, likewise, be balanced in its rest, its actions, and in its givings and regivings.

God's One Law of Balance must be as inexorably obeyed as that part of it which is known as gravity, must be inexorably obeyed.

To think what He IS, and what He KNOWS, into imaged-forms to manifest Him by motion, *God divides His still Magnetic Light into electric mate pairs. He extends these pairs which He has divided, to two measured Magnetic foci which He, likewise, extends to balance this division, and multiplies their power of control over action to manifest His Omnipotence in the ratio of their extension.*

The two Magnetic foci, which are Magnetic poles in man's knowing, are extended from His fulcrum stillness, along wave paths of His eternal stillness, to control their manifestation of His imaginings from within all creating bodies, and to balance their separateness from without.

He, likewise, divides His Omnipresence by centering every particle of creating matter in His whole limitless universe with the still Light of His own PRESENCE. From each center where He thus stands, He reaches out his right arm to one pole, and His left arm to the other one, to form a shaft around which each separate unit of all Creation must move to manifest the cycles of His thinking. In this manner each Mind-centered body in all Creation has at its command, and as its inheritance, all-power and all-knowledge to draw upon in the measure of man's desire, and in the measure of the awareness of God's centering Presence within each unit.

Aeons pass before aught but sensation—then instinct—makes bodies aware of the centering Light of that Divine Presence. Long ages pass before thinking and knowing transcend sensation and instinct. When thinking begins, knowing also begins. More long ages pass before God's ultimate Creation—MAN—knows of his divinity through full comprehension of the whisperings of the Inner Silent Voice which forever says to him: "What I am you also are."

II

God, the One Knower, becomes three by His imagining. The still Light of the Knower, and the moving lights of His thinking, are the Trinity which God is in all things in this universe.

God, the One Father-Mother divides His sexlessness to extend

Fig. 19.

Fig. 19 illustrates the entire process of creating—and recreating matter. The Father-Mother electrically divides His ONENESS into pairs of father and mother bodies. The electrical strain of separation is equalled by an opposite strain, or tension, of desire for ONENESS. Unity thus attained is repeated in similar electrically dividing and uniting pulsations, forever.

ANOTHER WAY OF STATING THE SAME THING.

God, the Creator, divides His one white Light by extending its ONENESS into electrical tensions of vibrating red and blue pairs. The tensions of this electrical division are equalled by a desire for unity, which is attained *at the point of white incandescence* in matter. Unity thus attained is repeated forever by the same dividing, uniting process of electrical action-reaction pulsations. Reproduction cannot take place until the red and the blue lights of sex-divided motion are voided in the still White Light of the Creator. Man alone, of all Creation, ever knows of his Omniscience.

father and mother bodies from His Oneness. *The one desire of these separated male and female particles, or masses, is to unite to void their separateness. Upon this formula God's electric universe of motion is founded.*

The sexless Father-Mother Creator is One. His extended sex-conditioned, male and female bodies are the completion of His Trinity.

Rest and action are three. Space and matter are three. Equilibrium and motion are three. Dimensions and pressures are three. The heartbeat of the universe, and yours, are three. Likewise, its breathings and yours, its temperatures and yours, and all things else of the universe, and you, are three.

The swinging of the pendulum is three, as the spectrum and the fulcrum and lever, also, are three.

The cathode is one—but its extended pairs of anodes in the electric current of man, and of space, adds up to three.

Silence is one—but sound springs from silence when its divided moving pair collide—so sound is three, and its vibrations in sequences of rest and action, are also three.

God is ONE in all CAUSE—but in all EFFECT He is three. And all that are three are nine—for all that are three are multiplied by three in this visible cube dominated universe of three dimensions.

III

God decentrates His electric thinking to imagine idea. He then concentrates to form a moving body-image of idea.

Mind-*decentration* and electrical-*depolarization* are one. Mind-*concentration* and electrical-*compression* are, likewise, one. *Generoactivity, compression* and *polarization* are also one. They constitute the life principle. *Radioactivity, expansion* and *depolarization* are *one*. They constitute the death principle.

Man's Mind and God's Mind are ONE. Man's thinking and God's thinking are ONE.

Man decentrates to conceive idea and concentrates to create a body in the image of his idea, exactly as God does.

God's thinking is electrically expressed by extension from a point of rest in space to a compressed point of rest in matter. It then electrically expands to a point of rest in space to disappear into its Source. This is God's way of dividing rest with action. It is also man's way during all of his brief cycles—but when the rest interval for the longest cycle comes he calls it death and the end, for man does not yet know that God's ways, and man's, are ONE. Nor does he yet know that he cannot die.

Man charges his batteries that way to give them life. He discharges them to void their life—then recharges them. All bodies are batteries. Growing bodies are charging batteries. Nature's bodies are charged batteries. Discharging batteries are dying batteries. Discharged batteries are like unto dead bodies. There is no life in them because there is no motion. Nature forever recharges her bodies—beginning her charging in their seed and discharging back to seed.

Life is motion. Death is rest. Each is fulcrum of the other. There is no death in Nature, save man's belief in death.

God's Magnetic Light is eternal life. God's thinking is eternal life in action, divided by rest. Life in matter is but a pulsing simulation of eternal Life in God—the ONE.

IV

God's thinking is universal. His actions spring from His thinking, therefore, God's actions are universal. Thoughts do not take place just *here,* or *there,* where they begin. They are everywhere, and their beginning and ending are one.

Actions, likewise, are as Omnipresent as their Source in Mind-thinking. That which happens anywhere happens everywhere in this universe of naught but Mind-extension.

Idea has no extension, but idea, divided by imagining, extends into an imaged infinity, and repeats its divisions like unto the infinity which the kaleidoscope repeats, and multiplies its imaginings as it repeats.

When God thinks at any one point of rest in His universe, that point becomes the center of an invisible cube of White Magnetic Light. From there it is harmonically repeated as cube centers throughout His Cosmic Kaleidoscope, at the rate of several hundred billions of cyclic pulsations every second. Their speed of extension into this three dimensional illusion is about 186,400 miles per second.

Radical expansion of beginning points compress motion into cube planes of rest in space. These are reflecting mirrors of Magnetic Light which project God's thought-imaged forms onto His universal screen of space to simulate a reality of existence where not anything is, not even the motion which so convincingly *seems* to be there.

Invisible cubes of Magnetic Light, and of zero curvature, are the boundings of wave-fields within which the curved universe of reflected spherical forms are projected to constitute this electric thought-wave universe of complex illusion.

God begins each electric thought-wave at a point of His White Light at the intersection of the three inner planes of the cube, which are at right angles to each other. This point of beginning is the wave fulcrum. It is also the point of idea-conception in Mind. It is the centering eye of the inert gas of the elements which springs from that plane. It is, likewise, the cathode center of man's electric current and the beginning of the wave-shaft which extends two ways to divide the red half of the spectrum from the blue, to create separated father and mother bodies. Here also is where time and all other *dimensions* begin, as well as all other *effects,* such as life, compression, polarization and heat. Here also is where all depolarizing *dimensions and effects*

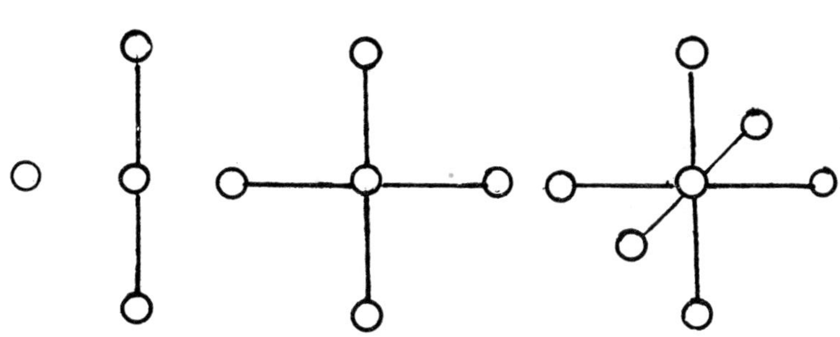

FIG. 20. FIG. 21. FIG. 22.

The nine zeros which bound the three projection mirrors to cause the illusion of a three dimensional universe.

end and disappear into invisibility, silence, and cold stillness of space.

God creates three pairs of *disunited* male and female bodies upon His extending octave wave shaft—and then a *united* fourth

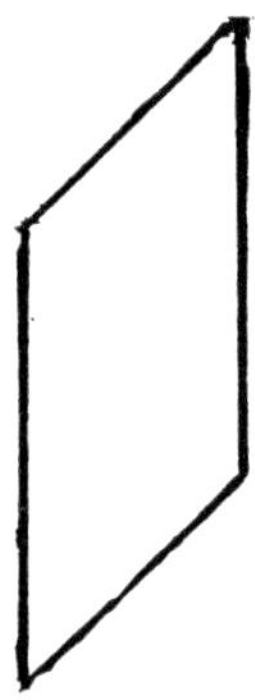

FIG. 23.

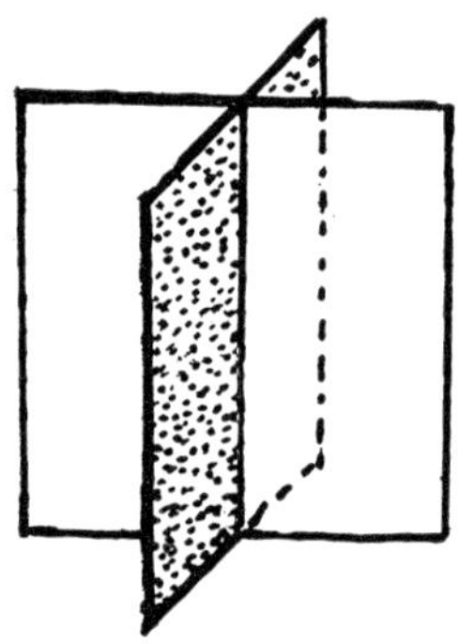

FIG. 24.

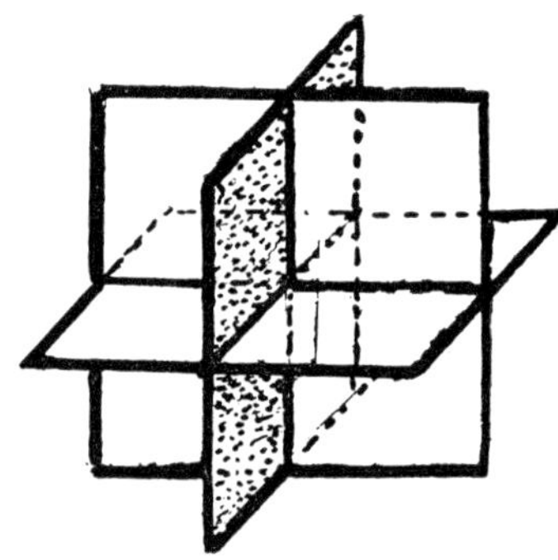

FIG. 25.

The three projection mirrors of the Cosmic Cinema.

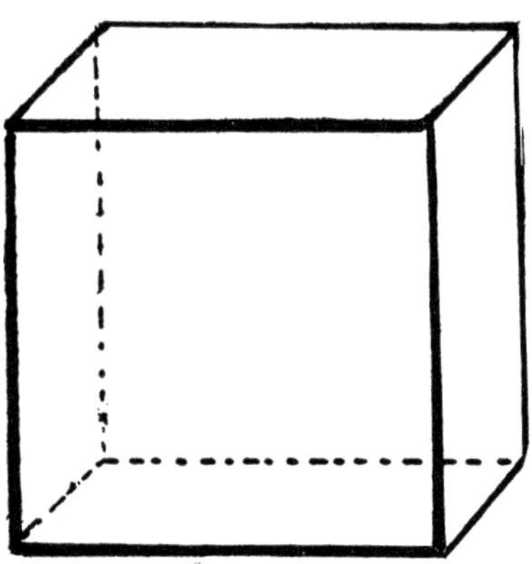

FIG. 26. The six mirrors which form the screen of space upon which the cosmic drama of Creation is thrown.

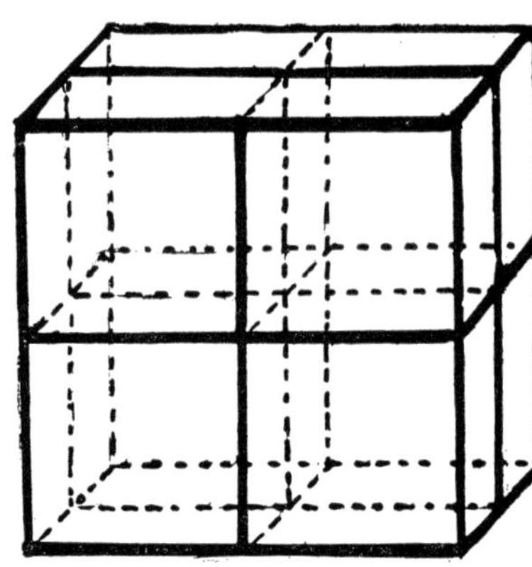

FIG. 27. The whole wave-field projection machine of nine Magnetic mirrors, which create the illusion of form and motion in this zero universe of electrically recorded Mind-imagining.

pair at His thought-wave amplitude. This united, balanced pair is an incandescent sphere. At that point in the wave, and that only, the disunited father and mother unite as ONE at its white

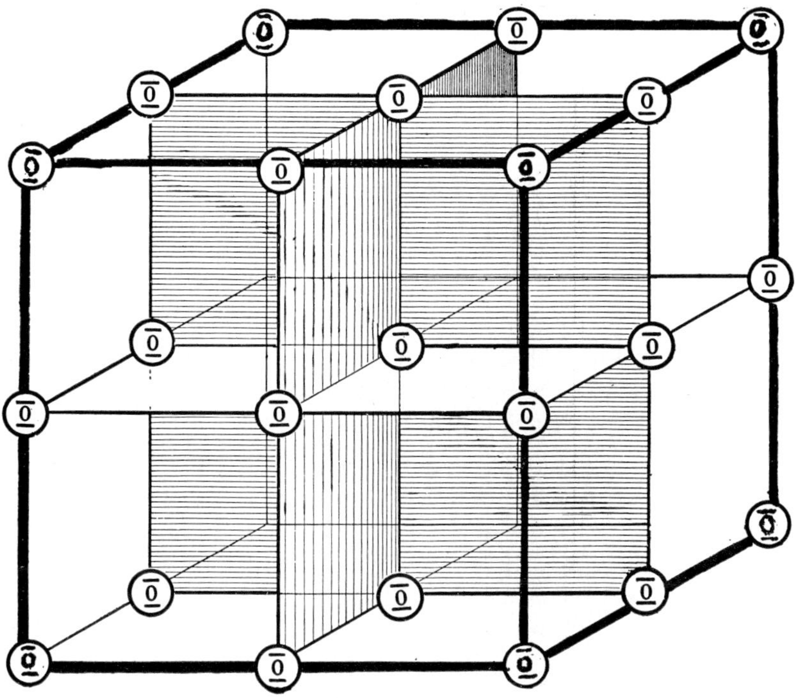

FIG. 28

Every polarized action-reaction counts up to nine—never more—never less. The wave octave formula for the elements of matter and the color spectrum is nine—being eight, centered by zero, as follows:

or

0 1 2 3 4 5 6 7 8 9 0

upon which our decimal system is founded.

1. The total count of equators is nine.
2. The eight corners of the cube, and its centering zero, equal nine.
3. The eight sections of the divided cube, centered by zero, are also nine.
4. The zeros upon each of the nine equators total nine, being eight centered by zero.

center. One of its hemispheres, however, is still the red light of the father and the other one is the blue light of the mother; for each is still extended from its centering Oneness.

V

God's Magnetic cube is three, multiplied by three. Its planes of zero curvature are nine, and its boundary angles are the eight corners and the centering one of the fulcrum Source. Cube wave-fields are the eight mirrors of Magnetic Light which project dimensioned and conditioned forms to all the universe from one wave-field to another throughout all Creation. Divided light opposes its division. Opposed pressures arise from resistance to this division. Curvature arises from resistance to two-way opposed motion. Resistance is gentle at cathode beginnings but multiplies its resistance with cyclonic fury at anode endings. Here is where gravity collisions of sex unions borns whirling, incandescent carbon, silicon or suns, according to the measure of Mind-desire exerted electrically at anode points.

Within the Magnetic cube of zero curvature a universe of opposed curved-pressures is born. Each curved pressure within it is a lens to multiply or divide, heat or freeze, compress or expand, solidify or vaporize all pressures of motion which pass through its concavity or convexity. In this manner God's curved universe of curved directions and curved cellular forms appear upon His Cosmic screen for an interval to simulate the many and the complex, and then disappear into His Oneness to rest for an interval between thought-pulsation frequencies. Thus the cube, which is the Oneness of all form, is imaged as the sphere in God's imaged universe. The cube and the sphere are one. *The sphere is an incandescent cube, and the cube is a frozen sphere.* The planes of the cube are nine, and their projections into the spectrum of the incandescent sphere are nine.

Frozen incandescent spheres of carbon become cubes. They image the cube of their cold wave-field. Their wave position is the only one of undivided balance in the wave, and all other positions are unbalanced because of their separateness, but balanced with an equally unbalanced mate. That is also why the very many moving particles in octave waves exhibit different qualities and transient changing forms to which man gives so many names, not knowing that one which he names *this* becomes *that* almost timelessly. That is why heating, moving forms are curved and cellular, while cold ones lose their curvature and reflect the planes of zero curvature of their cube Source in space. Water drops are cellular when warmed above their freezing points, but below that point they become hexagonal crystals to reflect their positions in their cube wave. Vast complexities of crystal forms thus arise from balanced and unbalanced matings, and from separateness as well, such as the distorted cubes of copper or sodium iodide, or the octahedrons and dodecahedrons of more dense elements farther removed from wave amplitudes.

VI

God's Mind centers all things, all minerals, all vegetables, all animals, and every cell which constitutes their bodies. He gives life and purposefulness to all things. Bodies acquire awareness of purpose only through electrical messages of command from Mind which centers them, for no body could otherwise move, survive or fulfill its purpose without being centered and polarized by Mind. Cells, glands, white or red corpuscles, hormones and other parts of bodies, must fulfill their purposes. In themselves they are helpless to move or act their parts in Nature's plan. Each part and each whole of all cell groups is centered by the Intelligence which centers the whole structure, whether ant, violet or man. Every creating particle of matter in the universe is a polarized Mind-extension. When each particle disappears to rest for each pulsation interval it withdraws within that Intelligence from which it extended as a patterned form of idea.

Man is the consummate image of God's imagining, just as carbon is the consummate element. *It is not a part of God's intent that carbon should have inner awareness of its divine origin and identity, but it is intended that man should have.* Mind and Soul centers every carbon crystal, never-the-less, or it could not fulfill its purpose as part of the patterned body of tree or man.

In all this vast universe there is naught but Mind and thought-motion. All motion is but an electric recording of the Mind-thought which centers it. It is also the record of the idea it simulates. In the idea is purpose of idea. In the electric recording, therefore, the mechanics of idea, thought and purposefulness are lodged, otherwise The Creator could not create.

God's universe is living, and is purposeful. Wherever there is motion, there also is God commanding His thought-forms to fulfill their purposes.

Know thou that God does not extend His Self into his moving universe, for the God-Light is still. Its stillness centers all things —and it, likewise, centers the *shafts* of all motion which turns around it, *shafts* which are levers of fulcrums and end at poles which measure extensions.

Naught exists but God. Man exists as ONE with God, but until he is aware of his Oneness he is but a thought-recording image of God's imagining. Some day he will know, however, for that is God's intent in creating Man. Know thou, therefore, that motion merely simulates God's knowing, and God's qualities, and the purposefulness of His divine drama of Creation. Simulations are not reality, however, nor do they exist. Imaginings come and go. They change and have dimension. God's imaginings are not God, however. The play cannot be the Playwright.

VII

When God, the Father-Mother, divided the Light of His sexless Oneness into the red light of the father, and the blue light of the mother, He ordained that the father light must penetrate the light of the mother, and be forever within her womb to live, and without her womb to die. Thus it is that the red fires which center our father of earth, lie enfolded within the blue coolness of earth's crust, and the cooling blue oceans and atmosphere of the encircling mother womb. Thus it is that the compression of the mother womb generates heat to polarize and vitalize the father seed of life which is enfolded in that womb. That process of sex interchange between the blue and the red lights which beget life, continues to beget life to give back to its Source until the mother can no longer compress life into the father, and the father can no longer discharge heat into the mother to continue to beget father and mother bodies.

When this has come to pass both expand. The earth emerged from the surrounding womb of the sun to cool and thus beget a father within her pregnant womb, to continue God's one process of creating bodies until both father and mother slowly depolarize by expansion of both, and both continue their journey into the Magnetic cold from which they emerged.

Mother and father reverse their spectrum positions, however. The womb of the mother is on the inside and the father surrounds it by a ring, such as one sees in the Lyra Nebula. One can also see the birth of a new star in the very center of that great black cathode hole which the mother womb is. That is the way that God turns the anodes of His thinking inside out to rest, and outside in to again become anodes. God's process of creating bodies through sex interchange is based upon the sex urge of the divided color spectrum of light to void its color divisions and become the White Light of rest from which its tensions were extended.

God is ONE—at eternal rest. Creation is TWO in the perpetual tensions of motion. The divided two in action desire rest in Oneness. They find rest by interchange, but lose it as they find it until they can interchange no more. A long interval of rest in "death" then follows but it is only an interval. It is just one black gap of the many rests between actions of God's Cosmic cinema, which simulates the Idea of Creation which He has imagined into seeming being. Life in Mind is eternally existent. Life in matter eternally repeats its *simulations* of existence.

VIII

God is ONE. His Oneness is manifested in all things. His White Light must be manifested in His universal image. The incandescent white light of a sun center manifests the Oneness of its Source in the Magnetic White Light of Mind.

Man is the consummate manifestation of God's imagining of His very Self. God's image in man is not yet complete. The time will come to every man when the Light which is God will be known in every man. That White Light of God centers every man. Few there are, or have ever been who know that Light within them, but all men must eventually know that Light as their spiritual natures unfold.

God's Oneness is imaged in the elements of matter. Carbon is the consummate element beyond which there is no possibility of extension, even as the cube or sphere cannot possibly be extended. The suns of the heavens are incandescent carbon. The still White Light of Mind centers carbon and unites its male-female pair as ONE. All other elements are in carbon and are incandescent in God's suns which are the seed for His universal garden of living things. All things are always carbon when incandescent, no matter how they may be divided for enfoldment in the womb of creating things to become a violet, a willow twig or body of man. No matter what these may be their residue left over from white heat is carbon—only carbon, the ONE THING of all matter. All things in Nature spring from their patterned seed which is their concept in the White Light of Mind. The seed is the Oneness of the uncountable many which unfold *from it* and refold *into it*.

Likewise, all chemical elements of the octaves are red and blue lights which are projected from the pure White Light of their inert gases, which are their octave seed. They return to their invisible Oneness by radioactive emanations which are pure White incandescent, microscopic suns. Man calls them alpha, beta, gamma or helium rays as they emanate from tungsten, actinium, radium or uranium at almost the speed of light. Each of them is the seed for another body of its like kind, as suns are seed for all bodies.

Not one complex creation of Nature can lose its Oneness. God extended them from Him to be like unto Him in His image. The great oak has many parts which multiply in number as they unfold from the Oneness of their seed. Its countless parts take on dimension in length, breadth and thickness which were dimensionless in their seed. It weighs many tons and gives shade to many things while purposefully manifesting God in action. When it refolds within its seed to rest, however, all of these dimensions and multiplicity of parts disappear into the Oneness of its seed to regive to their Source that which had been given to the tree. The tree is not dead, however. What it has been it still is. No microscope of man can find one electron of that tree within its seed, for if one could find what is really there one would find God. The patterned tree is the image of Mind-imagining. Mind-Light projected that image into space to manifest Mind-imagining. Mind rested between its cycles of imagining and withdrew the image into its equilibrium. *Every unit of Creation unfolds and refolds in that life-death manner.* Nothing has happened to the Idea of the tree. It eternally exists and will again unfold into action and again become the imaged form of Mind-imagining in repeated sequences when conditions of electric pressures are favorable for its reappearance. Even though ten billion years pass, and this planet has journeyed beyond its present pressure conditions, that tree will reappear on Venus, then on Mercury, just as it long ago appeared and disappeared on Saturn and Jupiter.

IX

God thinks in electric pulsations which are recorded in motion as four pairs of rings which are compressed into spheres. Each cyclic pulsation is manifested by the projection of four concentric light rings in one plane from one point of Magnetic Mind-Light, in which the red half of the spectrum is on the outside of the rings and the blue half on the inside. These four rings are the seed of the octave wave and occupy that position in the wave known as the zero group of the elements, or inert gases.

The wave is created by dividing the four sexless rings of the inert gases into four pairs of oppositely sex-conditioned rings, and projecting them toward sex mates of adjoining wave-fields to find balance and unity in each other. God's concentrative thinking compresses these mate rings as they are projected. This is the generoactive (uphill flow of energy) principle which multiplies power and speed in the inverse ratio of the cube, as they are thus centripetally projected, until the red and blue pair of cyclonic vortices, thus resulting, collide at wave amplitudes, midway between the two zero cathodes from which they were projected. This is the manner in which unbalanced and separated sexed pairs are united into the Oneness of the two

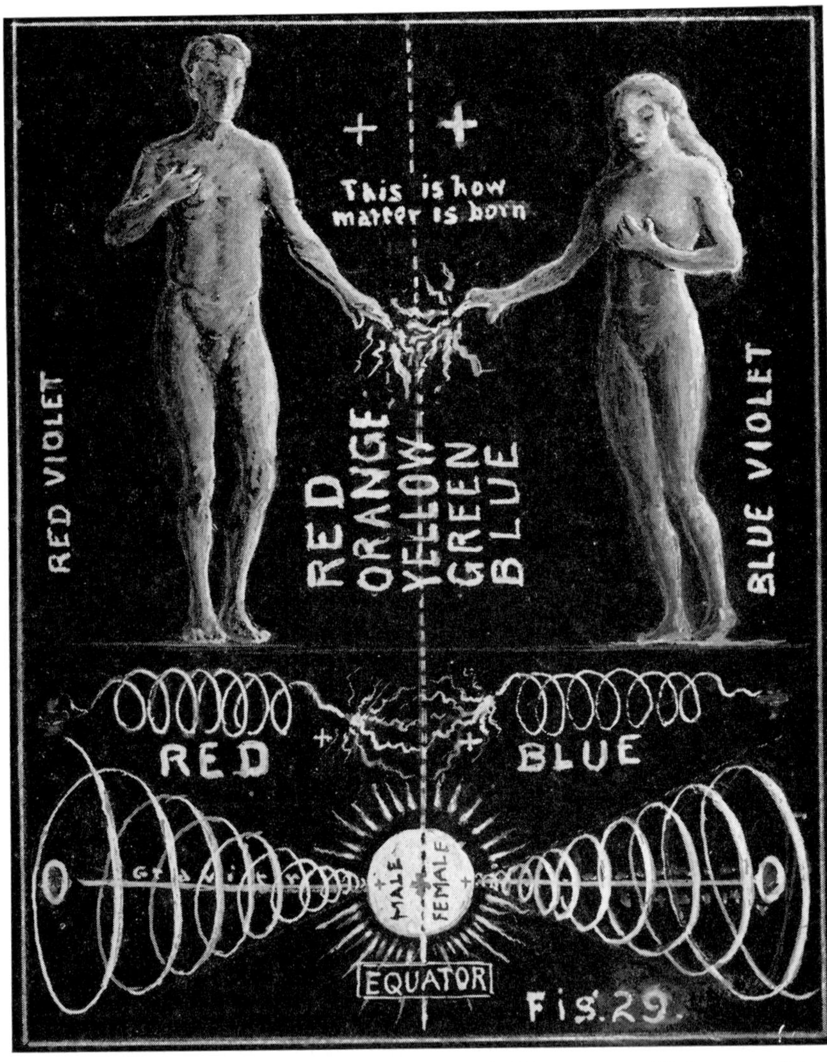

Nature creates sex-conditioned bodies by polarizing an equilibrium condition. Nature destroys bodies by depolarizing them. Nature repeats bodies through interchange between their opposite conditions. All bodies are eternally repeated.

balanced and equal hemispheres of spherical incandescent suns.

Within the four zero rings of the wave is the cathode mother womb of space which is seeking the outside to fulfill her office of borning the seed of the father. To aid this process the four pairs of projecting rings gradually close up their centering holes as the rings are compressed from their cone bases toward their apices, where the collision of sex-mating completes the closing in the incandescent sphere thus formed.

The wave shaft itself is the still God-Light of Mind which is like unto the still eye of the cyclone at the polarized end of the cyclone shaft. The true sphere thus formed marks the maturity of the imaged-form which is completed at the junction point of eight cubes where the octave wave ends at amplitude. This is where microscopic suns of almost timeless duration are formed. Here also is where majestic suns of huge mass and durations of billions of ages in duration are also formed, to function as seed for borning of planets and lesser forms of God's imaginings. From that point of maturity of such flaming carbon masses as our sun, God's decentrative thinking causes these suns to project rings from the plane of their equators in series of four which, likewise, compress into planets in series of four. These, likewise, forever project rings in series of four until all that God gave in light rings of His electric pulsings are regiven to the four of their zero group as refolding records of that which has been unfolded.

Within this process lies the mystery of the seed and its growth as its pattern unfolds, and the record of the pattern as it refolds. And as suns throw off rings they oblate in like ratio and the holes come again as one can see in the Dumbbell nebula which was once a mighty sun, or in thousands of other ring formations throughout our heavens. See figures 30, 31, and 32.

God projects His thought-recording rings from the cathode zero, in pairs along the *wave shaft* in planes of 90 degrees from it. Likewise, He matures His thought-imaged form so that its equator is in a plane which is 90 degrees from the *wave axis*.

The creator's atomic systems do not begin in the wave in elements which are centered by holes. They begin only when cen-

FIG. 30. Illustrating the death principle in all matter. Spheres are created by compressing rings to form them. Spheres are then disintegrated by projecting rings from them until a black hole surrounds their axis of rotation and they gradually expand until they eventually disappear into their cathodes.

trifugal force has multiplied sufficiently to throw off spiral arms and rings from equators of matured suns. Until that time all forming elements are centered by the gradually lessening cathode holes of the mother womb until the invisible patterned *seed* of the inert gas becomes the visible patterned body form of the seed image. During this entire cycle the still Magnetic Light of the centering cathode is projecting rings outwardly from it in pairs around a wave-shaft, to manifest the Love principle of giving, while the compressed bodies thus formed are exploding outwardly, likewise, to manifest the principle of re-giving.

The result may better be pictured by the tornado which every action in Nature emulates. All electrical motion in the octave wave turns spirally around centering, still Magnetic shafts, just as cyclones do. All electrical compression begins by violent expansion to create a condition which borns its opposite, as all opposites in Nature do. This effect is the answer to the electrical engineer's question which asks WHY his electric current is only at the surface of his wires and cables, and not all through them,

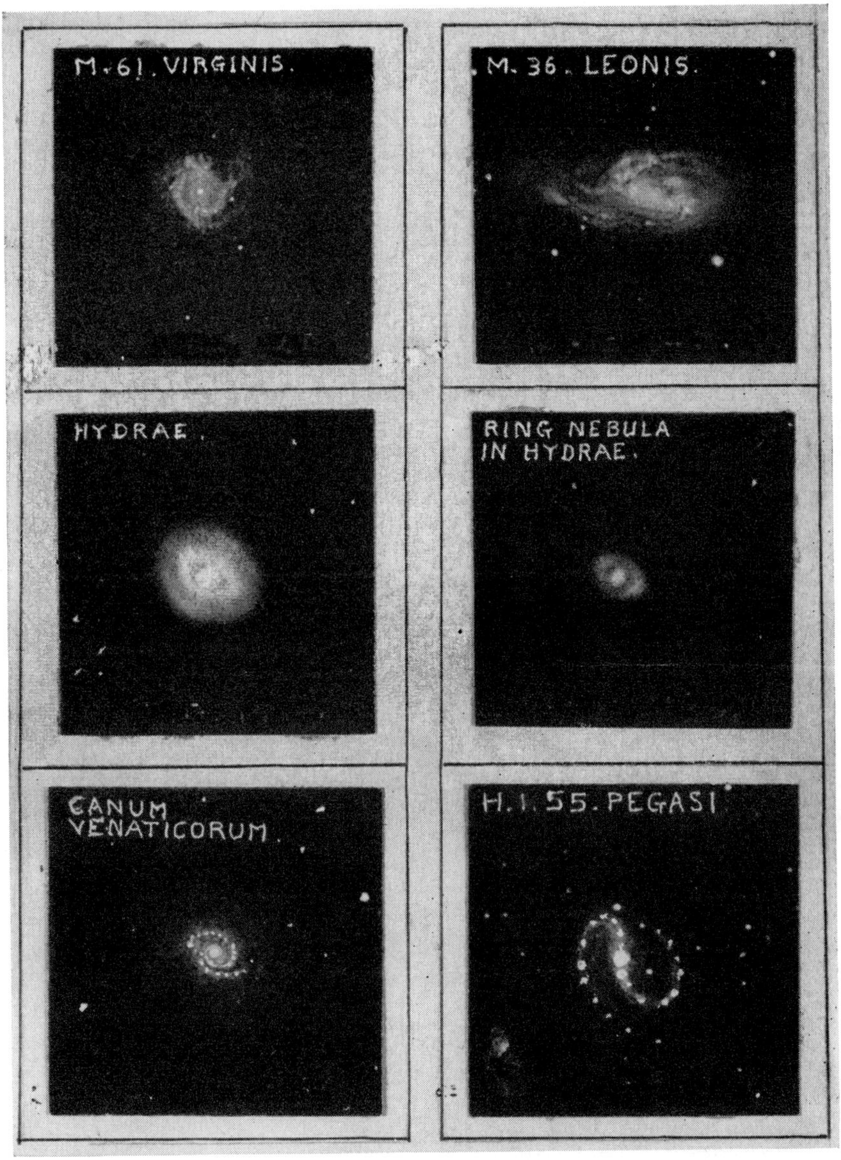

FIG. 31. All matter, from microscopic cells to majestic suns "unwind" spirally, and centrifugally by ejecting rings from the planes of their equators.

FIG. 32. The centering shaft, around which suns rotate, again compresses another sun into its vortex to repeat the process until all its mass disappears as rings. Lyra and Hydrae nebulae are good examples in the stellar system and Saturn well exemplifies how planets in our solar system, disappear into their zero.

except at the points where loops of force count his impulse frequencies for him. If it were possible for him to slice an electrical current into sections he would find that each section would be a ring whirling around a still center, except at points where collisions occur between the two lights of the spectrum, which are so forcibly projected from cathode zeros.

Thus it is that gravity is seemingly created to control the compressive force of Nature and the regiving reaction of the creation of an expansive force. Expansion is the result of MIND-DESIRE TO GIVE by its outward explosive effect from its zero cathode. The re-giving of compressed energy, likewise, is an outward explosive effect, as electric compression ceases, and the result of that effect is to born its opposite and leave great black holes within compressed masses as they return to the zero of their Magnetic stillness.

Thus it is that God's Law of Love is manifested in every action-reaction of Nature. *After long aeons of man's hard experiences in learning how to manifest the Love principle in his dealings with other men, he will some day know his own Oneness with God, and find happiness and peace which can only come to him by having made that supreme discovery of his own divinity.*

CONCLUSION

We have thus described the manner in which all creating bodies "emerge from space" for an interval to fulfill their destined purposes, and are then "swallowed up by space" to rest for an interval before again emerging to continue the fulfillment of their purposes. God's one desire to think action, and rest from action, in sequential intervals, is thus fulfilled. Likewise, His one action of giving for regiving to manifest His Love, is also fulfilled, together with His one motive for seemingly dividing His unity into two desires for unity.

These are the qualities of God's imagining which He projects upon the screen of His vast space in such rapid sequences of changing patterns that the senses of man are deceived by seeing *motion* where no motion is. Likewise, that which he sees as life in living bodies and death in dying bodies, is but simulated motion to simulate life. This universe of seeming motion is but an electric recording of Mind-imaginings. God's one purpose, which is to think what He IS, and what He KNOWS into His own image, is thus fulfilled. All of His Creation is for the fulfilling of that one ideal. Man is the consummate ideal of His

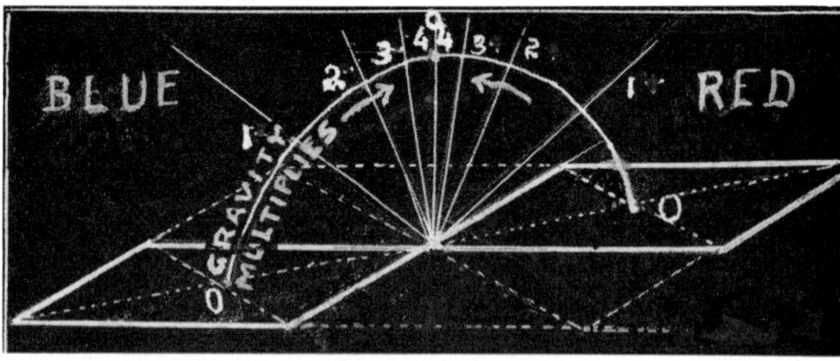

FIG. 33. Basis for structure of atomic, solar and stellar systems.

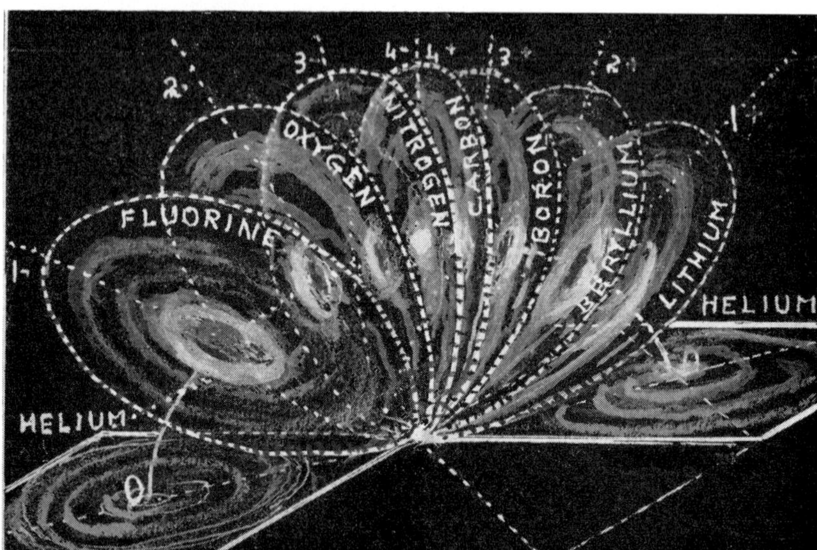

FIG. 34. Gyroscopic equators which unwind suns into rings by electric compression to rewind into stellar systems.

purpose, for man alone, of all His Creation, can become aware of his Oneness with His Creator.

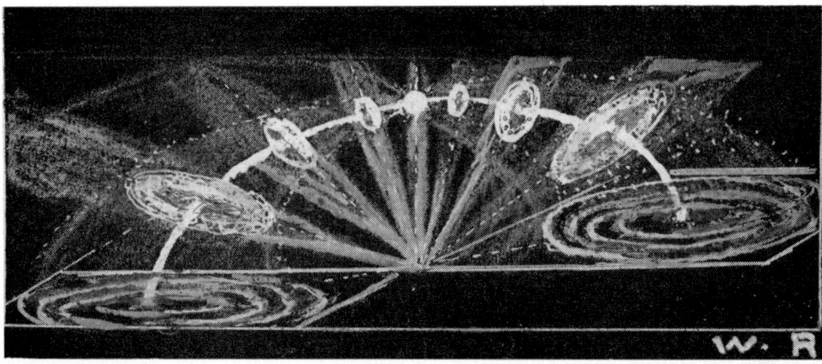

FIG. 35. Illustrating principle of construction of matter by the projection of polar rings from cathodes to collide as anodes.

Man himself, however, is but still in the making. For long ages he has been unaware of his Self because of building his body. The dawn of that conscious awareness has at last come to man, though that stage also is in its early beginnings. He is still so new in his spiritual unfolding that it is difficult for him to forget his fight for body survival in his jungle days. He has begun to listen to the Silent Voice within him, however, and is gradually becoming aware of the Light of His Source.

Some there are, however, who hear that Voice with greater understanding than others, but few who hear with complete understanding. These have become more illumined with the Light of Mind and their thinking has so far transcended their sensing that much of God's Omniscience is already theirs.

Such a man was Jesus, the Nazarene, who fully knew His Oneness with God. In Jesus God had fulfilled His complete desire for creating Man in His own image and likeness. Jesus gave to man that which He was commanded to give for man's uplift toward the Light, but man was not able to bear that which Jesus told him. Man crucified Jesus for thus claiming His divinity and for trying to convince men that they were like unto God. Man still crucifies all who come to transform him from the pagan and barbarian ways which still dominate man's relation to man.

God is a patient Father-Mother of man, however. The ages of time consumed in creating man in His own image mean nothing to Him. In this respect God says to man: "All men will come to Me in due time, but theirs is the agony of awaiting."

CHAPTER VI

The True Nature of Electricity and Gravitation

I

EVERY EFFECT of motion, whatsoever, in this universe is an electric effect. Electric EFFECTS are multiple projections from One Magnetic Light CAUSE. As there is no other force which creates this miracle of Creation there can be no theory of motion, or of construction of matter, which is outside of the electric process. This entire electric process is completely demonstrated on just one cycle of the electric current. *One cycle is the centripetal journey of a pair of divided units from their zero cathode to its multiplied amplitude, where pairs are united, and the return centrifugal journey to their cathode for rest.*

That is all there is to the creative process. That process is the same whether in microcosmic or macrocosmic scale. There is not one process for atoms and another for stellar systems. That same process is repeated in every action-reaction of Nature, whether it be the lifting of an eyelash or the explosion of tons of dynamite —or whether it be one's breath cycle, or casting a fishline into the sea. Our first step is to depict that simple process which has no variation in Nature, whether its vibratory frequencies are six billion frequencies per second or one frequency in fifty billion years.

Nowhere in the electric current of Nature do we find any justification for assuming that a material nucleus centers the spiral coils which constitute the universe. Instead of that we find the opposite. Nature so stoutly resists having the holes in her coils filled with density that she generates a terrific heat to demonstrate that resistance. That is how suns are made. Resistance to compression generates a terrific heat, which cools as suns expand to hole-centered rings, such as shown in the Dumbbell Nebula, Fig. 30.

The same thing will happen in your laboratory if you wind coils too densely. Every electrical engineer knows that he is limited in the amount of current he can send through a coil by the size of the hole he leaves in it. If a physicist or electrical engineer places a solid core at the center of his coils it will heat and polarize, but not strengthen the current. If he places iron filings there they would not remain in the center, nor could they be compelled to remain there. They would either line the inside of the coil or extend to its poles. If they were tied there they would become incandescent if the current was strong enough. *That proves that the core of electric motion is intensely resistant to any attempt to disturb its zero rest condition.* It also proves that Nature never creates material nuclei for her spiral cyclones, which constitute the electric current or electric waves of Nature. It also proves by the very fact of the heat which is generated there, that it *resists* instead of *attracts*, for heat is the basis of all outward explosion.

The entirety of this electric universe of motion is expressed in spirals. Every spiral in Nature is centered by a hole. Every spiral is a continuity of rings. All Nature is made up of spiral sections, or rings. The coils of the laboratory are spirals. Every section of a laboratory coil is a ring with a hole in it. Even if one puts a steel rod in the core to polarize its ends there is still a hole of *stillness* within the rod, for the electric current spirals the surface of the rod only.

There is no precedent in Nature, whatsoever, to justify Rutherford or Bohr in assuming that the atom is centered by a material nucleus. Likewise, Nature gives no precedent for concentric shells of satellites revolving around primaries in eccentric orbits of many planes. Satellites become eccentric, and their planes of revolution vary quite markedly from the equatorial planes of their primaries, but they do not start that way. They begin as rings thrown off from equators and gradually become off-center and off-plane. The three rings of Saturn, in Fig. 32, illustrate this beginning. The orbits of all planets and suns are rings with holes in them. The outer satellites, which have wound up from rings, illustrate the progression of rings on their journey to disappearance as motion. Anyone can prove by the following very simple experiments that there is no justification for the assumption of a nucleus in an atom, which is held together by some mysterious Cosmic "glue"—the descriptive name given to it by eager searchers:

1. Wind one hundred feet of copper wire into a circle and pass a current through it, leaving a large hole in that ring. The moment you bend the wire you also make the loops of force, which surround it, come closer together within the hole, than outside of it. In other words you make them radiate from a center instead of being parallel to each other.

Bear in mind that the loops of force which are spinning around the wire are creating potential. There is no potential at all in the core of the wire. That wire is an axis of gravity around which electric potential is being multiplied. That potential is being multiplied by fast motion around its stillness. In the gravity center of that 100 foot coil you will not feel an easily measurable amount of heat resistance to the compression being acted upon it.

2. Now wind that 100 feet of wire into eight rings. Those widely separated loops of force, which spin around eight wires instead of one, are reaching outward radially toward cold, and are reaching inward radially toward heat at the center of gravity, which centers your coil as the eye of a cyclone centers its similarly spinning loops of force.

The inner rims of rings, which constitute those loops of force, are coming closer together, which multiplies heat, potential and density, because of the outward explosive resistance to that compression. Now test the hole for heat and you can quite easily measure it. Also, test it for potential and you will find that density increases in the direction of the center, for resistance increases in that direction. Now, test it for polarity and you will find that a very thin needle on the end of a thread will seek two points of stillness where gravity is being divided, and potential is being multiplied at 90 degree angles from its axis.

3. Then do the same thing with sixteen rings, then thirty-two. You still have a hole in the middle. You find that heat, potential, polarity and density constantly increases as you make your spinning radial rings smaller, by winding your coil with more and smaller turns.

If you now put a steel rod through those rings you will find that you need two needles, instead of one, when you had but one ring. You will find that as you extend your rings you also extend potential. Gravity immediately divides into pairs when potential extends from its Magnetic zero, and electric potential multiplies by so doing. Polarity is that effect in Nature which is caused by dividing one point of Magnetic stillness into pairs of still points. A zero shaft is thus produced around which electricity spins its rings to create the electric potential, which is so casually referred to as "matter."

You will also find that the steel rod has no current at its center, but only at its surface, except where it is concentrated at its poles. If you now test it for density, potential and heat you will find that each increases in the direction of its center, which is exactly what you would find if you could test the sun in the same way.

4. To make a more comprehensive test, and to also convince yourself that the needles are seeking stillness in the eye of the electric cyclone which you have created, instead of being attracted there by an inward pulling force, make a coil in the shape of two cones, base to base. If you now try your needles you will find that it is impossible to make them seek the center of any of its rings other than at its very extremities. If you try to have one seek your central ring it will forcibly resist it. If you hold two at that point, just a half inch apart, you will find that each one will seemingly repel the other. They are not doing this, however. They are each seeking a point of stillness which is maximum in gravity resistance to the electric potential, which

has multiplied to its maximum power of compression at that point. Instead of being attracted there they are compressed there as they point out the direction of increasing density and electric potential. This is proof, also, that the effect upon the needles is an electric effect—not a Magnetic one.

If you are not sufficiently convinced by this experiment make two conical coils and put both apices together. If they could meet at a sharp point you would need but one needle to find one common center of gravity. If you now spread the cones apart and put some fine iron filings there, you will find that they will gather together in a ball, and your needle will point always in the direction of increasing density in that ball. You need but one needle now, for two will do the same thing.

Now pull your cones apart far enough to elongate your sphere into a spheroid. You can now use two needles, for you have again divided gravity.

For a last convincing experiment, take your needles to the holes in the bases of the coils. Turn the coils on end and lower one of the needles where the potential is strong. You will find that the needle will follow the gravity shaft and point directly toward the eye of the vortice, as indicated in the cyclone diagrams in figures 51 and 52.

These familiar experiments should convince you that there can be no nucleus to a mass which is being compressed into an unwilling and resistant center. Where is there any evidence of a material group nuclei in these atoms which you are thus creating?

It is necessary that you fully understand that all matter is explosive and that every body of matter thus compressed against such resistance is desirous of exploding. Cold alone imprisons matter into dense solids, liquids or gases.

The Scientific American recently published an article by Robert Hofstader in which he said that individual units of matter might prove to be merely rings spinning around an empty point, with maximum density at that point and a decrease of density from the inside out.

The following drawing and the caption under it, are reproduced from this article.

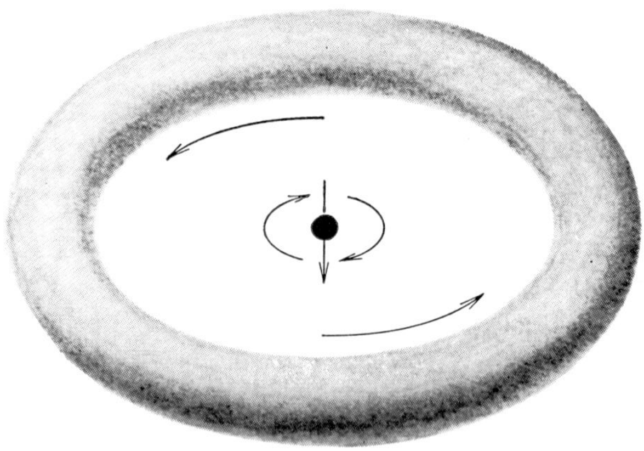

FIG. 35. "MESON THEORY suggests that the proton may actually consist of a spinning 'bare nucleon' which is essentially a point, surrounded part of the time by a rotating meson cloud."

You can see by the above that some modern thinkers are rising above the illusions practiced upon them by Nature.

5. Try this experiment also. Pump air into a deflated tire. The more you pump the more you compress. At first it is like having one turn of wire around a big hole. The more you pump the greater the potential and heat you create. The pressure you put into it desires to escape. It, therefore, forms a dense ring inside the tire but your first thrusts of pressure seek the outside of the rim in their desire to escape. The more you compress, the greater the density, heat and potential. Its increase is in the inward direction. *Bear in mind the fact that you are compressing from the outside inward. You are not attracting it from the inside. The inside is always explosive.*

One of the greatest causes which led to the assumption of an atomic nucleus is the familiar spiral nebulae of the heavens. As you examine them in these pages you will see a central sun in all of them, around which suns and their planets are revolving like a pinwheel. These spiral nebulae are all dying stellar systems, which do not become systems until after the collision of octave pairs of rings at wave amplitudes. In other words an atom, such as hydrogen, oxygen, nitrogen, aluminum or sodium is a single particle—not a system. The radar principle of Nature determines its tonal wave position. Each of these are rings around holes. Carbon is a system, for it is a united pair which has reached maturity and must now begin to die by throwing off the rings which made it what it was. Every sun you see is flaming carbon. True it is, that it centers a system and appears to be its nucleus, but it is not a nucleus which is holding its system, or itself, together. It is a Cosmic powder keg which is doing its best to explode. It cannot do this all at once for the cold of the zero universe will not let it. It cools gradually and dies as it cools. There is no force of any nature which holds it together by an inward pull. Nature does not attract, nor does it repel. It compresses within a vacuum and the vacuum restores its normality. Electricity creates tensions which the universal vacuum releases from tension.

Figures 36 to 40 illustrate a basic fact of the electric current. As long as the two dividing poles extend away from each other as they leave their cathode to collide with an approaching pair, which has been extended from the next cathode, speed of motion multiplies centripetally and potential multiplies with it. This is what is meant by the Einstein Equasion of 1905, for that fast motion does multiply potential.

The moment the collision takes place and the two poles become one in a spherical body, that moment there is a reversal of the entire pressure principle. The acquirement of polar unity in one gravity center touches off a reversal lever in Nature. The Cosmic clock spring has been fully wound. Now it must unwind. Instead of the collision of two approaching poles the united one divides into two, which continue right through to their opposite cathodes. It is as though two apices of cones meet to create one center of gravity, then push on through each other until they bore a hole right through the compressed sun or other body. See Fig. 30. In so doing a body which has been compressed by increasingly fast centripetal motion is expanded by increasingly fast centrifugal motion. As potential increases by fast revolution

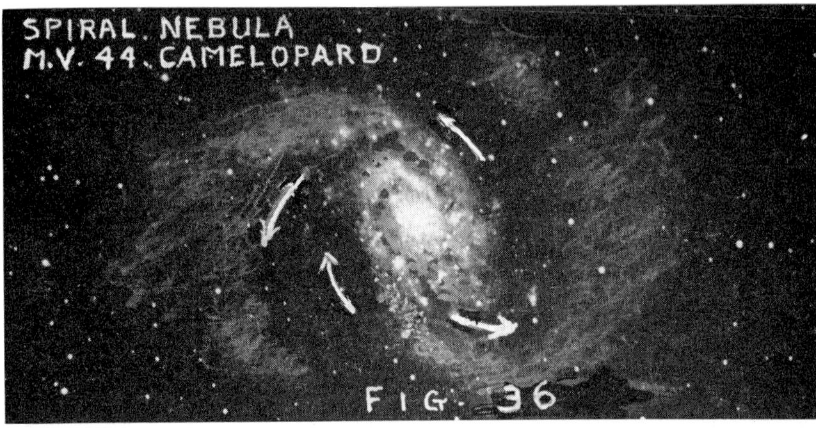

The supposition that gravity is a force which pulls inward from within is, basically, the cause of the misconception of the nucleal atom. Instead of holding matter together the nucleal suns of these systems have accumulated enough heat, by multiplying it centripetally, to explode them centrifugally.

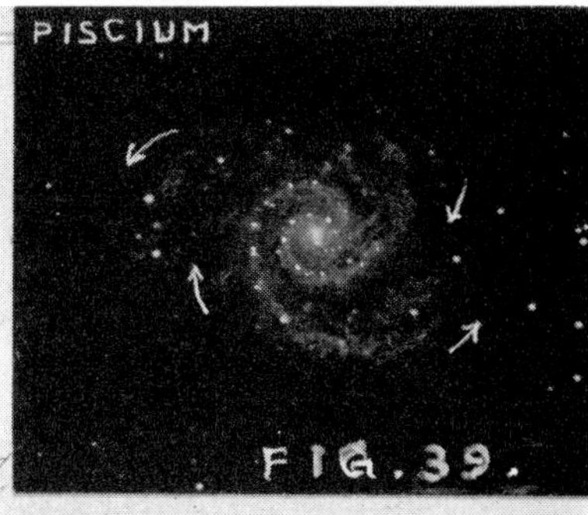

The nebulae on these two pages undoubtedly influenced Rutherford and Bohr in forming their conceptions of the nucleal atom. Lack of knowledge of the nature of gravitation and the true nature of electricity, caused these misconceptions.

around two *approaching* poles it, likewise, decreases by fast motion around the hole being made by receding poles. The following is a heretofore unknown basic principle of motion.

Everywhere in Nature life and growth are expressed by increasingly fast centripetal motion, and death and decay are expressed by increasingly fast centrifugal motion.

In observing the opposite directions in which the arrows in the nebulae diagrams point, do not be confused into thinking that the many masses are turning in opposite directions. Every unit in all of them turns in the one direction which is common to all Nature. The opposite directions of the arrows indicate direction of increase and decrease in electric potential, which is the way all things live and die.

Fig. 40 represents a nebula which has not yet reached maturity. It is still being compressed toward a perfect sphere. It is a united pair but has not "had any children" yet. An atomic, solar or stellar system is one which has reached maturity as a united pair, and has begun to propagate generations of "off-spring", which, in turn, have propagated generations of their own. All atomic systems are discharging, depolarizing, expanding systems. There are no polarizing, or charging systems in Nature. Depolarization alone produces multiple systems. It is because of this fact that the "expanding universe" theory deluded so many observers.

Figures 36, 37, 38 and 39 are old systems. Figures 38 and 39 are excellent examples of the regeneration of new solar systems along the expanding paths of dying suns. Their symmetry, and very noticeable balanced rhythm, are very indicative of normal disintegration.

II

THE HEAVY PRICE OF SENSE-PERCEPTION

It is right here that all observers, from Newton to Einstein, have been so grievously misled by the evidence of their senses. *All of them have believed in a force in Nature which pulls inward from within.* Out of their observations we have, in physics, a law covering the acceleration of gravity, and mathematics to give ratios of the power of matter to attract matter, and many other misconceptions. The very fact that a stone falls to the ground has built a belief that gravity attracts it. *Instead of that it is the density of electric potential which causes the stone to seek its like pressure condition in motion.*

The stone is densely wound up into a high potential, high melting point state. It belongs in an electric pressure which has caused that state of density, and it will seek it—*not because it is attracted to it*—but because the vortices of compressing motion are pulling it that way. Every potential state of pressure in this universe has its proper place in it, and it will seek the current of motion which takes it there. Raindrops will fall to seek equality of environment in the direction of gravity. Compression is the cause of that fall, *not gravity*. Water vapor will rise to seek equality of environment in the direction of space. When water vapor finds its equal pressure it will float weightlessly, for weight is but the measure of potential which is out of place. Matter which floats in equipotential environments is weightless in respect to any other matter in the universe, yet mathematicians spend much wasted time in calculating the weight of this weightless planet.

SUMMARY

Gravity exerts no force, whatsoever, to either attract or repel. Gravity is a shaft of Magnetic stillness which has been extended two ways from zero in the cathode of the electric current, around which electric motion can simulate the power—or energy—of its Magnetic zero center. Gravity is motionless and changeless. A gravity shaft is a two-way extension of a point in space where all the power in all the universe is existent, to a pair of points around which motion turns to balance the power extended from their centers. Electricity spins around these pairs of gravity points, as they extend into shafts, to *simulate* that energy by motion, but the simulation is not the reality. When each pair of gravity points meet as ONE, gravity unifies and dual polarity ceases. The Magnetic stillness of the cathode where division began is then repeated in matter.

CONCLUSION

The thoughts which the above is written to accentuate are: 1—that every effect in Nature which observers have attributed to gravity and magnetism should rightly be attributed to electric potential, and 2—that the center of any mass like the sun is a center of maximum heat and pressure which is a violently explosive condition, not an attractive one, for heat expands and violently desires to escape.

Tens of thousands of suns have exploded. They are called novas. Many of them are conquered by the cold of space and re-imprisoned. The greatest of these is known as the Crab Nebula. It is a very recent explosion but a very spectacular one. A careful study of its movements gives one a clear comprehension of that majestic battle which is continually going on between the cold of space and the heat of its resistance to its electrical division. These misleading effects in Nature have held scientific progress back for too long already. Their correction would be of incalculable value to present day progress.

III

The text books tell you that there are two kinds of electricity, a positive kind and a negative opposite. Let us show you how impossible and illogical this is, and how contrary it is to Nature's processes in the construction of matter.

We again repeat the fact that electricity does only one thing—it divides an equilibrium into EQUAL pairs and compresses them until they unite to create an explosive condition around a still point of gravity. When that purpose has been completed electricity slowly dies in all masses, until all motion has ceased in them.

It does not require two forces to compress anything. If you wish to compress air into your tire you compress it with a force exerted in one direction, which is inward from the outside. The greater the force you exert in that direction the greater the multiplication of resistant pressure within your tire, as compared with the pressure outside of it. If you open a valve the pressure within your tire will explode outward without the aid of another kind of electricity to help it escape from its bondage. It will seek its equilibrium level without another kind of force to help it. The universal vacuum is that universal equilibrium level. Any departure from that state in Nature's normalcy is a forced departure which causes a tension, or strain. The zero universe is without tension or strain. *All matter is compressed motion. All compressed motion is explosive.* It is in exactly the same condition as your tire which you have compressed into a strained, tense condition, from which it constantly exerts its own strength of desire to escape into the universal equilibrium. This desire for outward explosion is inherent in all matter. There is no desire in matter to hold itself together with other matter.

We refer you back to figures 6 and 7 for further study. Note that the zero one is free from tensions, and the other one is full of tensions from which they can be freed only by generating enough heat to begin their slow process of radiation, or decay.

It will help you to better comprehend our meaning if you stretch a piece of elastic from its normal equilibrium condition of rest. It takes force for you to stretch it but it will return to its normal condition without need of another kind of force. In stretching the elastic you have created strains and tensions of an abnormal condition. The zero universe is balanced. Everything in Nature which becomes unbalanced by the exertion of any force will eventually find balance in the Cosmic vacuum, which is the one normal condition of space. Is is also the CAUSE of all EFFECT and the SOURCE of all ENERGY. If you fully comprehend this, you can now solve that great mystery which great thinkers in science thought to be insoluble—the mystery as to how matter emerges from space and how space swallows it up again.

How was it that early investigators decided that there were two opposite kinds of electricity instead of the one kind which produces motion? It was because the two opposite conditions of living and dying—growing and decaying—heating and cooling—polarizing and depolarizing, and all other effects of motion, are expressed in seemingly opposite directions by seemingly opposite forces. *There are no opposite directions, or opposite forces, however. There are but divided sexes which exert the same force and in the same direction. The one force is compression and the one direction is spiral. That which seems to be two are one when they are united. They could not unite if they were persuing opposite directions, nor could they be one if they were opposites.* One's senses are very deceptive. They convincingly make one believe the very opposite of what they manifest.

Early investigators did not take into account the fact that motion is a cosmic abnormality which has been caused by a disturbance of stillness. The normal condition of this universe is a rest condition. Motion is a created effect which emerges from rest and returns to it. This universe of motion might be likened to a quiet pool into which a stone has been thrown. The normal quiet of the pool has been disturbed by a *force.* The normal quiet will return without the aid of force. There is no opposite force which causes the return to normal balance.

And so it is with life and death. They are two seemingly opposite effects which emerge from the Creator's zero universe as a disturbance of its vacuum by seemingly opposite pressures. So, also, are heat and sound. But all of these which emerge by the application of force, return to their normal rest condition without the aid of another kind of electric force, or a change in their one universal spiral direction.

Let us be sure that you understand this vital fundamental of Nature that has so grossly deceived the greatest minds of the centuries. We return to the tire which you compressed into a very much higher pressure than that of its environment. That pressure is held in the tire by a sealed casing, but it is very difficult to entirely seal it against slow leaking from some part of it.

Always remember that every action you perform causes motion—and motion is electric—and electricity moves spirally—and that spirals are always created in pairs. You cannot cut a section through any electric current anywhere without producing rings which spin around holes. Try it. Pass an electric current through an evacuated tube with sufficient air, or vapour in it to aid visibility, and you will see the rings which electricity creates.

You will see them as rings of light spinning around black holes. Those rings you see are divisions and extensions of the "ultimate particles" of Creation, for there is no other form in Nature than opening and closing rings. *They are the basis, and the substance of all form.* Whey they appear, matter appears. When they disappear, matter disappears, and all effects of matter with them, such as sound, color, heat, form, density and dimension. See Fig. 41.

Once again we repeat that this universe of motion is entirely electric, and there is no power, or quality, in electricity to pull inward from within. Again we say—electricity does but one thing—it compresses to divide into pairs for the purpose of creating a dense pressure condition known as electric potential. This is done against the resistance of the universal vacuum, which finally conquers every effort of electricity to simulate cohesion.

The entire principle of the construction of matter is based upon surrounding an area of rest in the omnipresent vacuum with four rings, then in compressing the rings in divided and extended pairs until the holes are eliminated and flaming carbon suns occupy the holes. Suns then throw off rings in series of four until the holes return. Electricity causes the compression and the zero vacuum is the expression of Mind-energy which causes the expansion. This universe is a compression-expansion pump. One end of its piston is in the eternal vacuum and the other end is in the pressures of electric potential.

Remember, also, that every electric action, which is recorded in Nature, like the growth of a tree, or throwing a stone in water, produces rings with holes in them. The young tree starts that way, as a tube, but closes its holes by compression, to become a solid, and every solid is a series of ring layers which eventually open to let "space" in gradually, until space becomes all and the tree disappears entirely into it. Cut your own body into sections and you will find it is composed of rings around holes—your chest—your skull—your bones, arteries, heart, windpipe, nerves and every cell of your body. Electricity works that way. It tries to close up its holes, but very few out of millions of effects suc-

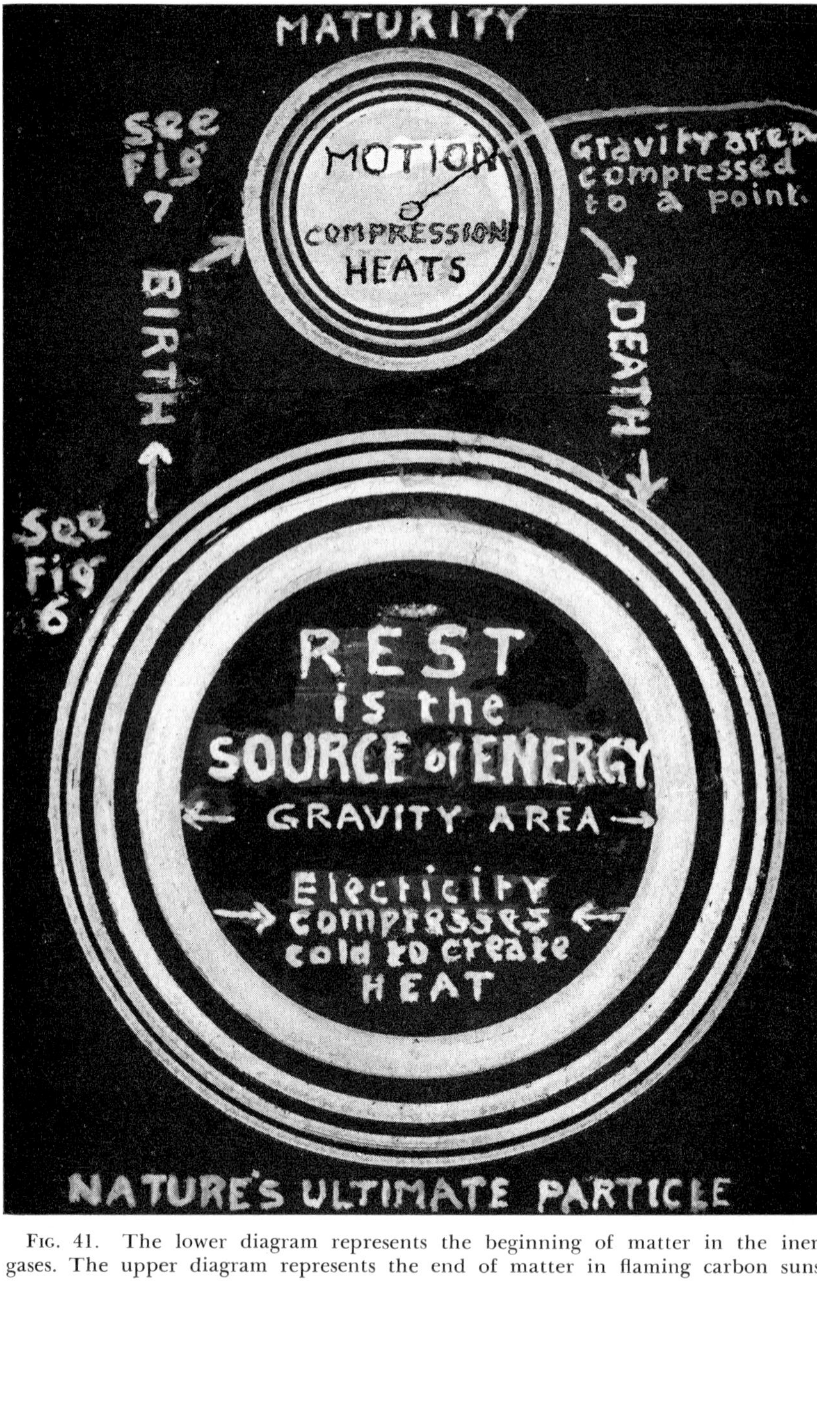

FIG. 41. The lower diagram represents the beginning of matter in the inert gases. The upper diagram represents the end of matter in flaming carbon suns.

ceed in doing so. Organic life has not one example of body building which has succeeded in becoming a solid, not even the ivory of an elephant's tusk. It is centered by a hole, and its cells are porous.

All Nature, everywhere, cries out its protest to such an unnatural and impossible condition as the nucleal atom. Nature is cellular and cells are rings in sections. Also every cell in Nature is a product of the union of four pairs of rings. Nature also cries out its protest against such a concept as that of a cosmic "glue" of some mysterious nature, which supposedly holds the atom together from within a nucleus. *Atoms are held together only by pressure from the outside and sealed from the outside by cold.* The inside of every atomic mass is a heat generator and heat consumer. Atomic units have cold centers, but combined atomic masses have relatively hot centers, according to their purpose and position. Its cells must be electrically conditioned to maintain that heat, but no matter what the temperature is in any cell it tends to *expand* it—not hold it together. Instead of being a glue it is an explosive.

IV

It will be well to examine the reasoning process which led observers and research workers to determine that there must be two kinds of electricity. The idea of a universal vacuum never entered into the thinking or reasoning of early observers of EFFECTS of motion. Lacking that concept they lacked their first essential premise—the basic fact that Creation is a product of MIND. Secondarily they conceived heat to be the reality which caused motion, and considered cold to be just less heat, instead of cold being the eternal uncreated CAUSE from which heat was created as an effect. Instead of heat being the cause of motion the reverse is the fact. Heat is the result of motion. Both are abnormal conditions in this universe of rest.

Now it so happens that the early concept of electricity provided for a one-way direction which compressed to multiply potential, which simutaneously multiplied heat. The opposite flow was theoretical but necessary, for there are two poles, and two opposite conditions to every electrical effect. There is but one direction to motion, however. This is a two-way opening and closing universe, both of which are expressed in unchanging one-way direction.

The sex divided condition did not occur to these early observers. Such an idea never entered their thinking. To them sex was a function of organic life and most distinctly not a part of the atomic life of inorganic matter. Nor did the spectrum red and blue divisions ever become a part of their consideration. Nor did the idea of tensions ever enter into their thinking in relation to electricity, the tension of the spectrum division which desired unity by the disappearance of the colors of motion into the Magnetic White Light of universal stillness, or the tensions of sex division of Father-Motherhood into father and mother bodies which desired unity in sexlessness. Nor did they take into consideration that the two opposites of compression and expansion coincided with concentration and decentration—growth and decay—life and death—or the polarization and depolarization principle.

Then, in the turn of the century, Rutherford and Bohr conceived the idea of atomic construction as being based upon the firm belief in the Coulomb law, which says that matter attracts oppositely "charged" electrical matter, and repels similarly "charged" matter. Nothing could seem more convincing, for one pole of a magnet seemingly "attracted" its opposite and "repelled" its like. It never occurred to them that males and females do not unite with their own sexes. If the sex idea, in relation to electricity, had occurred to them there would never have been a Coulomb law, nor a nucleus in an atom. Nothing could be more convincing to sense-reasoning than the very self-evident fact that there must be two kinds of electricity—even if the electric current did run but one way, but with an unproved suspicion that it ran two ways.

Much confusion has also been caused by the fact that the two poles extended in opposite directions from their dividing cathode and approached each other from opposite directions to form an anode. This led to the belief in the *opposite directions* of the two kinds of electricity, one of which was assumed to *attract* and the other to *repel*. The answer to this is that polarity is not motion. *It is the stillness of gravity which centers motion.* It is the omnipresent zero. It, therefore, has no direction. Motion is confined only to the electric rings which are forever encircling gravity. Those rings have but one direction in their turnings around their omnipresent gravity controls, but they cause two separate effects. One of these effects is centripetal and the other is centrifugal. Because of these two life-death effects the universe may be described as a two-way opening and closing universe of but one direction.

Some new and important discoveries were made which proved that there were two kinds of electricity. We will recite one of these discoveries which clinched the belief in two kinds of electricity, one kind for each way. You can read it for yourself more fully, if you choose, in the Encyclopedia Britannica, under the heading of "Electronic Tube." It reads as follows: "T. Edison observed the passage of electric current in one direction from a hot filament to a cold metallic plate in an evacuated enclosure, as if negative particles were emitted from the filament."

Let us see how this fact was misinterpreted. There was nothing in this experiment which warranted the attention it received. The electric current which Edison observed was the radiation from a heated condition seeking an equilibrium. It did not need to be in an evacuated tube. It is the same effect which your hand feels from the rays of the sun, or your body feels from a hot stove in a room. It has always been known as radiant energy. Radiation creates electric current. So does generation. One compresses, the other expands, but it is the same electric current. Any movement of any kind creates a two-way electric current, which traverses the whole universe—even the movement of your little finger.

A hot iron at one end of a room and a block of ice at the other end will cause an electrical current to flow both ways, until the iron, the ice and the room are all equal in temperature. That will also happen if you put two hot irons at opposite ends of a room. Warm water rising into cold space creates an electric current, but that does not mean that there is one kind of electricity to make it rise, and another to make it fall. *One might as reasonably say that there are two kinds of water, the kind which rises and the kind which falls.*

Such a claim is like saying that there is one kind of electricity which makes a man live and another kind to make him die. Electricity is centripetal when it multiplies its potential by increasing its speed, and it is centrifugal when it decreases its potential, also by multiplying its speed.

Electricity is motion—any kind, or stage of motion. There cannot be two kinds of electricity. There is but one kind of electricity but that one kind multiplies its power to compress in the first half of its cycle and divides that power in the second half. Much confusion is caused by mistaking speed for high potential, and vise versa. For this reason it is well to exemplify our meaning by the following example: To create matter by the compression of four pairs of rings, projected from cathodes, the speed of the current around its shaft of gravity multiplies constantly and *volume decreases as speed increases,* until the collision takes place. Simultaneously the speed of rotation of units in the current slows down until the completed mass, after the collision, is its minimum. From that point on every effect is in reverse. Revolution around the shaft of gravity of each separate unit of the mass, including projections from it like planets and satellites, become increasingly slower while rotation of each mass upon its own shaft is increasingly faster. In this case speed results in a lowering of potential and a vast increase in volume.

To exemplify: Mercury and the planets beyond it, revolve centrifugally around their primary shaft of gravity in the sun with ever increasing slowness, while the speed of rotation upon their own shafts increases so greatly that they are rapidly disintegrated by their own increasing centrifugal speed. This is a characteristic of every electrical effect, whatsoever. It is the principle which we call life and growth, which reverses itself to become death and decay.

Likewise, there are not two kinds of motion. Motion is the same kind, whether fast or slow. Air is the same air, whether hot or cold. Motion makes believe it is many things and many substances, but these are but many conditions of the same thing. *In all this universe we have but two basic fundamentals—rest and motion.* Together these two are the pulsating cinema which constitutes both Creator and Creation. As the human intellect unfolds it gradually sees the universe thus simply, and gradually becomes less confused as man becomes more aware of the reality of Mind and less dependent upon motion.

V

An electrically charged body is still a charged body until the last vestige of motion leaves it. Your car battery may be discharged to the last mile, but as long as it can produce motion it is positively charged. You might properly say that it is approaching a negative, or voided condition, but one cannot properly say that a discharging body is negatively "charged" under any condition. To use such terms as negative electricity or negative charge, is equivalent to using such terms as silent sound or dead life. The word negative means to negate—to void—to empty or become minus. To charge means to be added to. It is not logical or proper to use a term such as negative charge, which literally means added to by subtracting.

Electricity is the force used by the Creator to create bodies. Bodies are electrical disturbances in a vacuum. Electricity causes those disturbances.

The principle of electronics is a miniature example of what electricity does in a vacuum. A vacuum tube, with no electricity dividing it, is the normal rest condition of the universe. See Fig. 6. Electricity emerges *from* the omnipresent universe and disappears into it, as light appears and disappears in a vacuum tube. Electricity ceases to be electricity when it loses its power to move. When we speak of electric expansion it is in the sense that it is weakening, just as we say of a man who still lives, that he is dying. Electricity does not expand, nor is there a kind of electricity which does expand. It merely weakens by becoming lower in potential and speed.

When an electronic tube is evacuated it means that the electric commotion and tensions which were in it have been pumped out of it, leaving it in a zero state of rest. An electric current sent through it records within it the patterns of motion which caused the electric current. The vacuum condition insulates the new pattern of motion from any other patterns of motion, which would otherwise occupy the tube. That same pattern can be projected from that tube and repeat itself anywhere in the universe. That tube is a correct miniature zero universe of stillness which is creating a multiple universe of motion. If you will but study that effect you will gain much of Nature's processes. The more that science develops this principle of creating its own patterns by starting them from zero without interference from other pressures and patterns, and then multiplying them by gravity control within them, the more progress we will make in the practically new science of electronics.

The practice of electronics will grow to enormous proportions as electronic engineers begin to know more about electricity, and the relation of its pressures to the Magnetic universe which controls those pressures. That new science, though still in its infancy, has already made a marked difference in human lives. It will continue to make a still greater mark upon human destiny as it progresses. Its present handicap is its inefficient tubes and projection apparatus. Not any of these have yet been constructed upon the right principle decreed by Nature for maximum efficiency. That will come, however, as electronic engineers know more about the secrets of the wave, and the *radar principle as it relates to the geometry of space.* Man progresses only in the ratio in which he acquires new knowledge, and that comes slowly, for man is unwilling to discard old concepts for new.

This is not the place for a more extended treatise upon this subject, but we would recommend that each of you become more fully familiar with the daily progress now made in electronics. The one great lesson that you can now learn from electronics without being technical, is the fact that whatever patterns you put into those tubes, in the way of sounds, pictures or movements, comes out of them in the same patterns as those you put into them. Let us fully explain our meaning in this respect. The sounds and motion you make do not go into the tube. It is not necessary to tell you that, for you know that nothing goes into the tube but an electric recording of the wave patterns, of light and sounds, of the events which waves record. Now here comes your lesson. You can hear and see what you put into that tube as many times as you wish to turn on the current. That means that electricity is the receiving and recording principle of Nature, while the zero universe is the broadcasting principle.

This brings us to a high point in this book where we should pause for a moment and give deep thought to the realization that the Zero invisible universe is the THINKING-MIND-ENERGY-SOURCE of Creation, and that all material bodies are but the recordings of MIND-THINKING.

Should not every serious thinker readjust his life to seeking values which lie within their Source, rather than to continue to seek them in their shadows?

In other words, electricity compresses motion into seeming form and effect. That is all it does, however. When that compressed motion is released, to produce a sound for example, the sound expands into the zero universe without the aid of any force, whatever. The fact that you can *hear* the sound is because it is electrically reborn by being re-compressed into the density of your body. Your electrically compressed body is a receiving station for reborning the dying sound, just as a cliffside is a receiving station for reborning the sound as an echo. An electronic tube, and a cliffside, are alike in this respect. That also means that all matter is but thought-recordings, for all matter is electric motion. It necessarily follows that your body is but an electric thought-record of your Mind-thinking, which is operating within the Cosmic vacuum tube just as long as you can keep it electrically vitalized sufficiently to create the image of you which your thinking is making.

That includes every creating thing, and its pattern, as being thought-recordings which act for awhile then refold into little electronic tubes, which we call *seeds,* for intervals of rest. When we turn the electric current on to those electronic tube seeds, we can see and hear the record all over again, whether it be ourselves, a tree or a solar system. Is not that a lesson in immortality?

Within the eternal electronic vacuum of Nature every thought and action that has ever taken place during all time, lies enfolded there in non-dimensional omnipresent space. That is one more lesson that the electronic tube might clarify for you—the lesson of dimension.

Some day, during your spiritual unfolding of innersensory perception through meditation, you will become aware of the fact that dimension is but an imagined effect. It might help you to take the first step in gaining this awareness if you will but realize that when you look upon a fifty mile expanse of land, building, people and events, you see it all within a dimensionless pinpoint of space within your eye. The whole universe is thus reduceable to the non-dimension of Mind-knowing, which has caused the imagined universe to be extended from its eternal zero, which never exceeds zero in CAUSE, but seems to exceed it by dimension, in EFFECT.

God did, in fact, create the universe in His image, but dimension is as much a part of His imaginings as form and events are a part of it.

Yes, the above is a wonderful lesson for one who is mystified by the vague idea of *immortality,* but a still greater lesson of *universality* awaits deep meditation by you upon the thoughts given above.

You have wondered much about the unfolding of a tree, or a human body, from its microscopic seed. Many years are consumed in that process after you have put the seed in your own yard to generate an electric current for again unfolding the record contained in that patterned electronic tube. The electronic tube seeds of Nature have their film patterns enfolded within them. Man has to carry the patterns for his cinema separately, also his projection machine—whereas Nature's projection machine is the womb of earth. It will always reproject the patterned microscopic image into a three dimensional form if the conditions for that projection are right ones.

Now give thought to the little electronic seed tube which is planted in Africa, ten thousand miles away from your yard. Within that microscopic seed a terrific conflict is being enfolded, a conflict between men, planes and guns. It is taking place NOW —this instant—and is being simultaneously unfolded from its recording seed into other seeds in your very room. You can see and hear what is happening ten thousand miles away. You can see the activities of their African setting of desert, palm trees and a broiling sun, right in your room. You can hear the agonized and frenzied shoutings of people who are ten thousand miles away from you. Likewise, 50,000,000 people like you can hear and see the same events from 50,000,000 different pin-points of space upon this planet. Does not that teach you a lesson in the universality and omnipresence of all things?

Where, therefore, is dimension, or time? If that event is happening within your room NOW as an unfolding from an electronic seed, and that dimensionless seed is unfolding in 50,000,000 rooms, how can you say that it is ten thousand miles, and weeks of time away from you? If you place a yardstick upon a mighty oak and measure its many dimensions, and then the oak withdraws those dimensions into its concept, as recorded in its seed, or in your eye, how can you say that the dimensions you measure with a yardstick and weighing scales, have any reality in a universe in which the reality you believe in disappears before your very eyes.

Where, then, is REALITY? Is it in the motion which forever appears and disappears in the motion picture universe every time the current is turned on to reproduce it from its invisible storehouse, or is it in the Idea, which is its eternal Source? How long will it be before man seeks that REALITY which is eternal *within him?* How long shall it be before he *even begins to comprehend* the meaning of that command to seek the kingdom of heaven which is within all men?

As we look out upon this vast world of strife, seeing fear in the soul of man instead of love and happiness, and seeing greed there also for worthless quantities of moving matter, and seeing also desires for body-sensation rather than Mind-inspiration, we can but conclude that man is still far from knowing that Light within him which One Man whom man crucified knew when He said: "I and My Father are ONE."

CHAPTER VII

Our Eternal Universe

I

THE GREAT TELESCOPES of man have revealed thousands of spiral nebulae in the heavens which give the impression of great pinwheels of fire that appear to be unwinding like huge clock springs, which throw out many stars like balls of fire in misty clouds of fire as they unwind. The spectroscopes of the laboratory have shown conclusively that these nebulae are all rushing away from each other with tremendous speeds. This fact has given rise to the assumption that the universe is running down like a clock, and that when all of these hot nebulae have expanded into cold space the universe will die of a cosmic disease called entropy, which means heat-death. The theory which gave rise to this belief is known as the expanding universe theory. This theory, likewise, arose from another theory which conceived the universe as having had its birth from some giant cataclysm untold ages ago, which caused a huge ball of fire to form. The theory assumed that this cosmic fireball is now splitting up into smaller fireballs and gradually wasting away toward disappearance into the nothingness from which they came.

The human senses are easily deceived by what they see when not compensated by knowledge of what is known but cannot be seen. All rivers run forever in only one direction to the sea. Our senses tell us that but we are not deceived by that effect into believing that rivers will die, because we know that they are forever replenished by mists and vapors which we do not so plainly see. *Our universe is as eternal as God is eternal. It cannot die for God cannot die.* It is true that all of these nebulae are rushing away from each other. They are on their way to disappearance into the zero from which they appeared. That is the way of all things in Nature. That is what is known as the centrifugal downhill flow of the expressed energy of Nature. Its uphill flow is the centripetal spiral of its beginnings. The eternal balance in this rhythmic universe divides these two expressions of Creation equally. As a matter of fact the *entirety of Creation is an uphill flow of expressed energy.* Its downhill flow does not require an expression of energy. Its downhill flow is its dissolution. Man can interrupt its downhill flow, however, and make it flow uphill again to multiply potential. Nature continually does just that. *Creation is an electric effect of compression. Compression multiplies to accumulate mass. Expansion divides to dissolve mass.* Compression is an effort which causes tensions in a vacuous condition. Expansion is that vacuous condition. Motion is always seeking a level. The level EXISTS. Motion simulates existence. It requires effort to divide a level into two levels, but the two become one without effort. Compression is always TWO but expansion is ONE, which the two eternally seek.

Life is an expression of interaction between two levels. It requires an effort to maintain the two, but the moment that effort ceases both levels seek the one. That is why life requires continuous effort, but no effort at all is required to die.

II

The expanding universe concept could be true if revised in the following manner. Instead of one great cataclysm which caused one great ball of heat and flame all at one time in prehistoric times, substitute for it billions of microscopic heat producing balls for every cubic millimeter of this whole universe. These billions upon billions of atomic units come to life and end their life in death, for a duration of probably ten billionth of a second, but they have an accumulative power in potential and mass which adds up into suns. These suns have one life-death frequency in fifty billion years. Time divides and extends as mass multiplies and compresses. Each of the billions of nebulae and suns in the heavens is one of those giant cataclysms which must divide itself and die a heat death. They split up into other fireballs which, in turn die a heat death. *All matter dies a centrifugally dividing heat death, but it also lives a centripetally multiplying heat life.* Heat is a condition which is necessary to the creation of life in inorganic matter and organic as well. But heat is also necessary to create death. Matter would never disappear if it could be kept cold enough. Matter can die and disappear only by accumulating enough heat to consume itself. The cathode beginning of matter is an immeasurable low degree of heat. The anode ending of matter is an immeasurably high degree of heat. Flame is the maximum consumer of matter. All matter is self-consuming but its creation is an effort of its Creator. The Creator *gives,* but that which He gives must be regiven. That is law, and law must be fulfilled. Fulfillment is automatic, however. It is self-motivated by explosion from within itself. *Flame is self-explosive but the heat which created the flame is a compressive effort.*

Creation is an effort of the energy of Mind-desire in its Creator. Mind-desire is expressed by the electric thinking process. With the beginning of thought comes the beginning of the heat of motion. Heat begins when thought begins. Heat concentrates when Mind concentrates. It demands an effort of Mind-energy to concentrate and heat. Electricity expresses that effort. It does not require an effort for Mind to stop thinking and rest. Nor does it require the exertion of an electric force to make hot bodies cool.

Every reaction is within its action. Expanding nebulae are the reactions of the actions which consummated them. They are the flame-consuming death end of Creation. They are quite visible to the senses of man, but the beginnings are multiplyings of that accumulation and are invisible to his sensing. Exponents of the expanding heat death universe sit by the dead and dying carcasses of Creation and see only the death ends of living things.

We refer you to Fig. 6. It represents the formless, changeless, timeless universe of Mind. Such a timeless universe can have no beginning. *God can have no beginning, nor can He die. Likewise, Mind-thinking can have no beginning nor ending. All matter in this entire universe is the record of God's thinking.* Matter is forever coming and going, for God is eternally thinking, and resting from thinking in wave frequencies. Electricity is forever recording the comings and goings, the lives and deaths, of God's pulsing thinking. As long as God thinks in pulsing sequencies electricity will record His thinking in the pulsing motion of matter. *You must know, therefore, that matter is pure thought.* This fact the world must someday know. Jeans, a scientist of great innersensory vision, prophesied that the world would someday find this out, but he was at a loss to tell the world WHY.

You must, also, know that every action in human life, or throughout Nature, is an attempt (or even an experiment) in manifesting the law. Such an attempt may be well out of balance and result in discomfort or catastrophy for man or Nature. The reaction, however, is a *fulfillment* of the law. The reaction restores balance always. The free will of man to create chaos, or of Nature to create a cyclone, are equal, but his free will, or Nature's, are limited to the action. The reactions belong to the zero universe of eternal, unchangeable expansion. Unbalanced actions of the transient can in no way affect the eternal, for *the eternal is forever omnipresent within the transient. Creation is, therefore, but a multiplicity of compressed pairs of moving units within an omnipresent expansion,* which are forever maintaining a great effort to continue their *compressed* conditions. The moment they cease to maintain the *two* levels thus required of them to live, they find rest in the *one* from which they seemingly divided themselves into two to manifest the one. *Compression multiplies energy expression.* Expansion takes care of itself, for maximum expansion is CAUSE. No energy is needed for matter to die. It is needed only to live.

Creation never began and will never end. Such a concept as the birth of the universe theory now accepted as fundamental belongs to archaic ages, not to this age. Creation is eternal. You are a unit of Creation doing that which is expected of you to do to manifest the Man Idea in action. You must, however, learn that you cannot forever remain as body in the universe of simulation, or make believe, which motion is. There are intervals in which you must be wholly Mind to learn your part in the drama of Creation. You must then reappear on its stage in many thousands of rehearsals until you manifest the divinity of the Man Idea instead of his flesh alone.

III

The time has come when we must now learn that we live in a Cosmic Electronic Vacuum Tube of invisibility, as Mental Beings who project actions from that vacuum condition to manifest our creative desires.

Look within a television vacuum tube for a moment. All you can see there are flashes of electric light which come and go. They *come* only because of a Mind-desire to manifest invisible IDEA into visible form. They *go* when the invisible IDEA has been expressed and the desire for further expression ceases for awhile. Ask yourself this question: Will those light flashes remain in that tube unless forced to remain there by a power outside of themselves? You have but one answer. Those light flashes have been created to manifest Mind-Idea in action. They have no energy of their own. They will cease when Mind ceases to desire manifestation by acion. What has happened within that vacuum? Mind-thinking has been electrically recorded. That is all that has happened. *That is all that Creation is. That is all that electricity is, and that is all it does.*

We will carry the example of the electric tube still further. Consider that tube as a miniature replica of the COSMIC VACUUM ZERO of the Creator's Mind. Now consider it as YOUR Mind. Now follow that up by realizing that there is no light in the tube, because you are not recording your thought-images in them. That is the only reason. The moment you connect the electric pulsations of your divided thinking into that vacuum it immediately begins to record your thoughts in light flashes which come and go as the light flashes of suns and stars of the Cosmic vacuum come and go like flashes of fireflies in the meadow. If you will but think this through in your quiet hours of innersensory perception you will then fully comprehend that the light of motion, which matter is, has but one desire—to escape from the bondage of compression which keeps it forever moving to *record* Idea instead of *resting within* Idea.

Every particle of matter in the universe, from microscopic particle to giant sun, desires to explode. The only way it can explode is to generate enough heat within it to explode instantly, or to decay slowly. The slow process of decay merely means that the whole mass cannot generate enough heat to explode the whole mass, but each particle can generate enough heat—according to the melting points of each—to explode it all part by part, over a period of time. The only way that matter can generate enough heat to die slowly, or quickly, is through the speed of compression. If you drive an airplane at 200 miles per hour it will not generate enough heat to die quickly, but if you could drive it at 4,500 miles per hour it would disappear in white hot flame in seven seconds. This fact was recently demonstrated in a wind tunnel experiment at Langley Field, Virginia. This effect and an electrical short circuit are identical except for the seven seconds time element.

The only thing which prevents the continual multiplication of heat is the uncreated eternal cold of the omnipresent vacuum which not only surrounds it, but interpenetrates its every part. Cold freezes matter in space to keep it from exploding, just as cold freezes matter in your deep freeze. Cold even freezes the flame particles of suns to imprison them within themselves. Every particle of matter in the universe retains its appearance as a gas, liquid or solid only because it is frozen into the ice of itself. A vast mass of the vacuum cold freezes our sun into its density and keeps it from exploding. Some suns generate heat beyond their melting points and explode as novas, but some of them are conquered by cold before they expand beyond the limits of their exhaustion. They are then subdued for awhile, until they generate enough heat to try it again. This happens many times to many stars before they become permanently subdued. Many, however, explode and disappear.

Compression alone creates density, but compression also creates the heat of resistance to tensions. Vacuous cold takes that heat away, however, and leaves only the ice of that heat to simulate substance. Iron, for example, is so closely compressed that it can generate 1,500 degrees of heat resistance to that compression, before the eternal cold subdues it. It desires to explode, however, and will do so if you help it by supplying 1,800 degrees from an acetylene torch. Everything on this planet will explode if it can generate enough heat to do so. It has always been trying to do this through the internal fires of compressive resistance, which sometimes reach the surface of the earth through volcanoes.

The earth has reached so far away into cold space from its sun that cold has quite completely conquered it, but man is now attempting to create conditions which might make it possible to explode. *A full realization of the fact that matter is not held together by nucleal attractive action from within, and the knowledge that our earth is held together only by freezing a crust of stone around its internal heat, should deter humans from helping the earth to explode as a whole, as it is continually exploding in every little part of it, which we call growth.* We mean by that, that when a small particle generates enough heat from the sun's rays to explode, the cold of its environment freezes it and causes it to refold into a cell of polarized matter. Gradually those cells take on patterns. Tens of thousands of forms of animal and vegetable life then appear. That is the way things grow, and the only way they die is that they are enabled to generate more heat than their normal needs. *Living bodies are produced by compression. Dying bodies are liberated by expansion.*

IV

We shall now build up the nature of the electric current to make its multiplying and dividing two-way principle more clearly understood. The very first principle to make clear is that electricity cannot run in a straight line. It always spirals around a hole in one direction to create its two-way effect. An electric current on a straight wire does not run through the wire in a straight line, it spirals around on its surface. There is always a hole in the center of the wire. A current strong enough to fill the whole wire would melt it. A still stronger current would vaporize it into gases.

Consider the electric current as expressed by the wave as shown in Fig. 45. In that case the wave spirals around its axis of gravity like a cone. You can demonstrate our meaning by taking a clock spring of closely twisted wire and pulling it apart. You will see it coiling around its axis of gravity like a spirally whirling cone. That is electricity at work which you see spirally whirling around its centering hole.

Now consider the wave shaft itself. Motion whirls around that centering hole in four pairs of cones to every cycle. That is where electricity multiplies its terrific power to compress matter in order to create the explosives of which this entire universe consists. The wave is REPEATED in cycles, but each wave cycle is a series of four conical pairs.

The first essential for you to completely understand is the fact that there is always a hole in all matter and in every part of an electric current, except where collisions occur between pairs to destroy them by breaking them up into spiral ring systems. The other essential fact is that cycles pulse in two-way compression-expansion sequencies to live and die. Then they must expand into the four ring formations of their cathodes to again live and die. It is an extremely interesting, and very simple process which you must thoroughly know, for there is nothing else beyond it to know. *Every action in Nature repeats that one formula, from your thinking, to your breathing. Because of its great basic import we will leave no stones unturned to have you fully comprehend that master key of wave mechanics and space geometry which could make a new world of science if properly applied, and a new destiny of your own commanding if you apply it understandingly to every decision which you must make.*

Helium, for example, is the inert gas record of the carbon octave. If you "electrocute" carbon by a million volt current the carbon will return to its cathode birthpalce as helium to again become carbon. Likewise, if your body is electrocuted it will return to its eternal record of you to born a new body of you.

There are nine of these inert gases in Nature, as you will see by the nine stringed harp of the universe. The only difference in their structure is that each consecutive inert gas, from one to nine, is smaller than its predecessor, for each octave is a multiplication of its predecessor. Chapter XI will more fully describe the inert gases as bridges between man and God.

The following series of diagrams constitute the entire constructive principle of the electric current, as it flows around a wire, not through it, and as it is expressed by waves. One cycle is the same as a million cycles, so we will confine our explanation to just one of them.

Explanation of Fig. 42. The upper diagram, marked A—A, represents the creation of an octave of tones, beginning with the inert gas of four motionless rings, centered by the stillness of universal energy, to act as cathodes at both ends. This pair of four rings divide and extend their pairs of four rings toward approaching mates. Electric compression causes the rings to begin to spin, then to become smaller and turn faster as they approach the plane of collision where all four pairs unite to form a sphere. This represents the centripetal half of the journey which charges, polarizes, heats and multiplies potential. These are the qualities necessary for increasingly vitalized life.

Between the cathodes is the sun of the atom or solar system thus created by the manifested energy which has been projected from its cathode Source, to simulate that energy by motion.

The central diagram, marked B—B, gives the octave names to the four pairs of tones of the carbon octave. The centering sun is marked carbon. Carbon is a united pair. It is a true sphere when hot, and true cube when cold. An equator marks the plane of union of the carbon pair. One hemisphere is on the red side of its wave, and the other hemisphere is on the blue side. The names of the elements are marked under each ring and their places in the spectrum are marked above them.

The lower diagram, marked C—C, represents their place in the tonal octave of the musical scale. The inert gas is the keynote of the electrical octave just as the note of C in the musical scale is the keynote for that octave. The keynote is omnipresent in all of the elements of matter as well as being omnipresent in the musical scale. By omnipresent we mean that it is in each note as well as its own tone. In music, for example, one is always conscious of

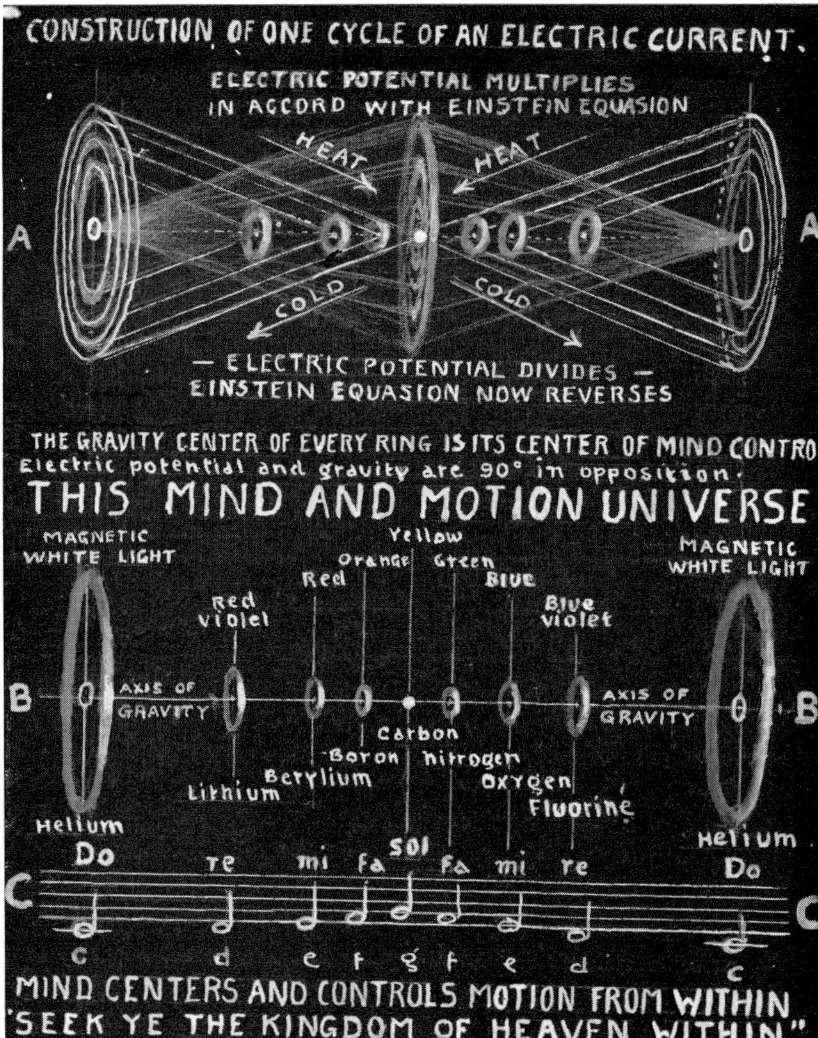

Fig. 42. Three examples of tonal rhythms of the electric current. Every cycle of an electric current is a complete octave of four pairs and an inert gas keynote. All motion in Nature is tonal and rhythmic. Its rhythms are geometrically and mathematically cube-based.

the presence of the keynote, no matter which one is being sounded, nor how many of them. In matter the inert gas is not so mind-consciously aware of its omnipresence but the sudden electrocution of any element, by passing a heavy current through it, will release all of the tones except the eternal keynote. By this process the area of gravity, represented by the hole, is gradually compressed out until the hole is closed up by the united pair and the very dense, hot, solid sphere is the product. The life half of the polarized body, which was created by an effort, comes to an end, and the death half, which requires no effort, begins.

This reversal from maximum compression to the beginning of expansion, and from heat to cold, and from charge to discharge and depolarization, is the most confusing of all phenomena in Nature to the physicist. It is the one thing all observers have missed. As we have mentioned before, even Einstein missed it in his Equasion of 1905, for the validity of that equasion ends right here and its reversal, which voids its validity, also begins right here. All observers throughout the centuries have failed to see that one direction of motion has produced a two-way effect, which is as applicable to a mathematical equasion as it is to a state of motion.

Observe carefully what now happens, *centripetal* force reaches its maximum and begins to die, and *centrifugal* force takes over. Look again at the top diagram in Fig. 42 and carefully note that the four rings of the inert gas are like the four rims of wheels placed within each other, with one common hub. Now note that these four rims become the hub when they are compressed and extended. Observe, also, that the extensions cause the appearance of cones as centripetal force winds the cone bases into a sun at the apex point of collision. Observe this effect also in figures 43 and 44.

Now for the reversal. If you look again at A—A in Fig. 42, you will note that the center of the inert gas is the point marked zero in both cathodes. These two points now become the apices of cones instead of the center of cone bases. The shaft of gravity which runs through from zero to zero now begins to open up at the center of the sun (or earth—or cell of your body) and cause

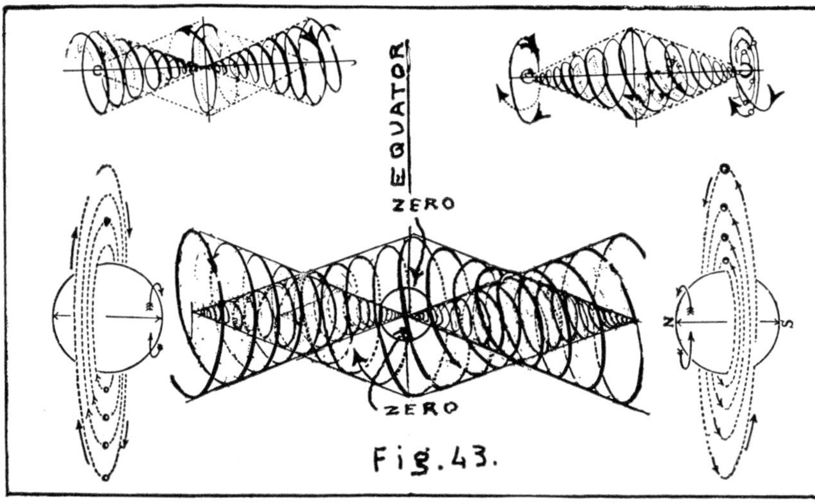

Complete life-death cycle as manifested in the electric current.

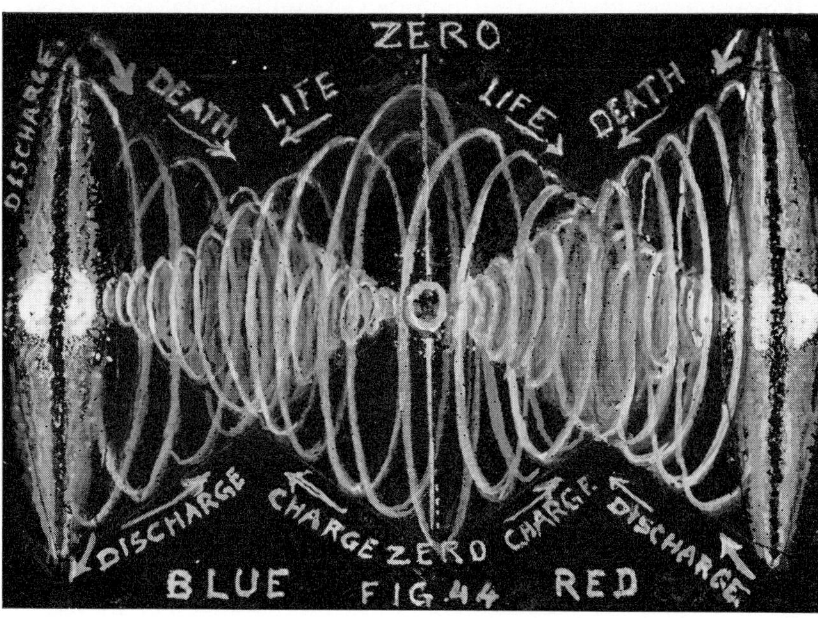

Complete life-death cycle as manifested in the heavens.

it to expand at its equator and flatten at its poles. Rings are then thrown off from the equator which are the bases of cones *whose apices are the zero centers of the cathodes.*

Now look again at these zero points in A—A of Fig. 42. Note where we have indicated cone apices there by faint white tones.

If you can now imagine yourself opening up that shaft of gravity in the center of any spherical cell, whether sun or carbon atom, so that the center of the sun becomes the center of cone bases, instead of being their apices, *you will have begun to comprehend a secret that the world has never yet known about the transition from life to death.*

Can you not now clearly see how centrifugal force of desire to return to the zero universe of rest, is unwinding that which has been compressed out of it, to let it come back into it? Can you not now fully comprehend the loops of force in an electric current, or giant nebula, as shown in figures 43 and 44?

If you now study the ring nebulae of the heavens you can see numerous examples of these dying suns. Many of them are shown in these pages. The Owl nebula is a very interesting example of an "unwinding" sun. It has a big hole clear through it. The Ring nebula in Lyra is also interesting to note because what was left of a sun regenerated at the center to repeat throwing off rings until nothing is left.

A complete explanation of this great pulsing movement, which motivates the universe, needs volumes instead of a few pages, but if you will follow carefully what is given herein you will have the essence of the profound mystery of the life-death movement of Creation, which the senses of man have not yet fathomed. To help visualize the process as a whole, however, instead of its parts, we print Fig. 45, which shows a whole carbon octave wave, rather than the electric current confined to a wire. The more you study these diagrams the more you will comprehend them. New thinking is not readily grasped in one reading. Meditation and repeated reading finally reaches your omniscience, *for your Mind really knows all things,* and needs but to be awakened to recollect that which it has always known. If you are not yet aware of this

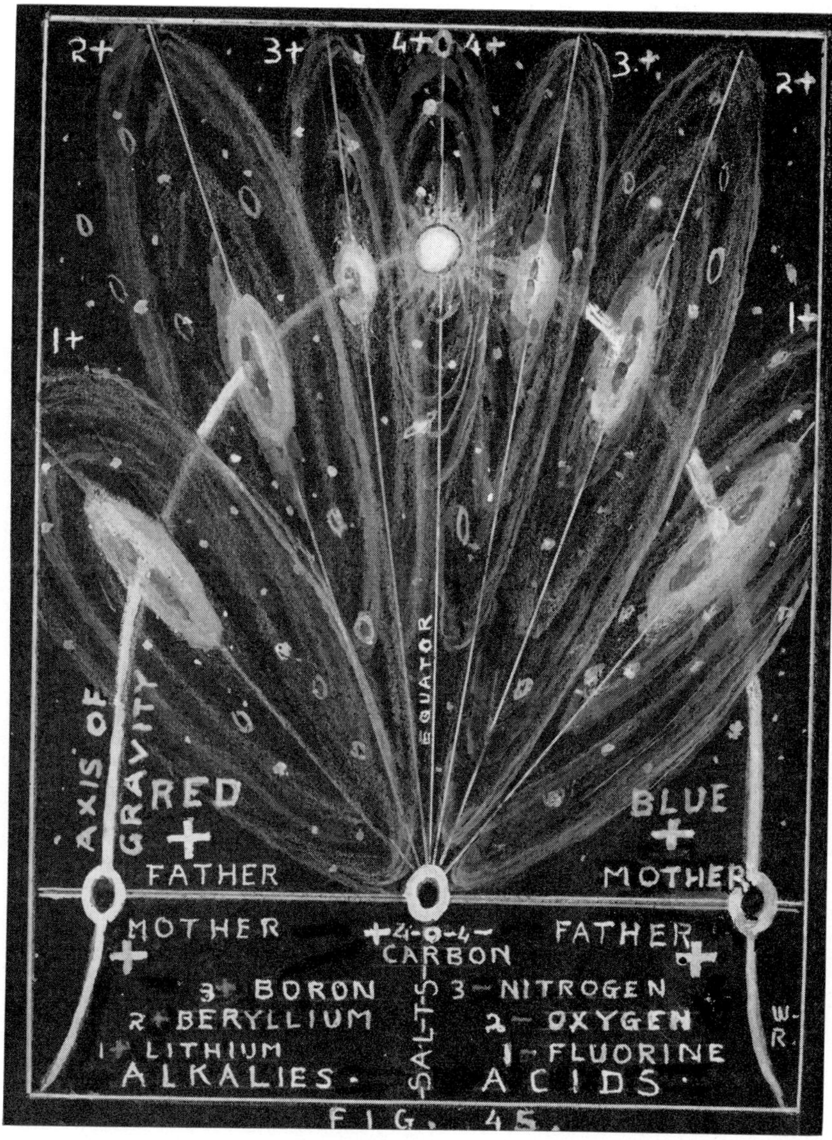

Complete octave of the carbon rings, as manifested in the wave.

fact you will be some day, and the more deeply you desire the coming of that glorious day, the more quickly it will come.

The most important and direct reason why mankind should comprehend the way the Creator works in this respect, is because this book is being written to demonstrate that man cannot violate God's orderly, rhythmic processes of Nature without paying a price which is equal in measure to the violation. Atomic fission is a violent abnormalcy for the rhythm of environment necessary for man. This we must show electrically, chemically and in other ways which man understands by laboratory work, for morals in Nature do not have a metaphysical basis, nor even an ethical one. The behaviour of all creating things is determined by their Source in the zero universe. These are the things which man who believes in free will to do as he chooses, must know. For this reason we must convincingly show him how close to God man is *without having known it.* That we must do in language and diagram which are familiar, and acceptable to science. A metaphysical or emotional language used with moral persuasion, in a plea to cease this terrific violation, would have no effect upon industry, government or science whose need for a life-saving fuel is great, unless accompanied with scientific conviction that such procedure is in defiance of Cosmic law, which will not allow such defiance. This volume is necessarily too brief for expanded detail. Since we must, therefore, keep to the point at issue, we will save much space by telling you briefly right here what the following pages are to reveal in respect to vital knowledge, which no man on earth today even suspects, and (to our knowledge) only one man of all the past ages has fully known and told to a world, which crucified Him because of it. Bear in mind that we shall not deal in abstractions, nor theory, nor affirmations, even though such affirmations may be admitted truths.

The time must eventually come when man must know where God is at every moment in respect to himself. He must completely understand how God controls every action and desire of all living things, from man to fungus cell, or from galaxy to electron, during their entire journeys from their beginnings in Him to their endings in Him. These are the unknown things which unfolding man must know, before he can begin to manifest his own divine inheritance as an omniscient and omnipotent Being:

1. The invisible universe is in absolute control of the visible. The invisible universe dominates and controls all motion by magnetic division into cube wave-fields of zero curvature, beyond which no moving thing can pass. It can, however, repeat itself in neighboring wave-fields, but always in reverse, as mirrors reverse. The geometry of the zero universe is based upon the cube, and cube sections. Their planes are of zero curvature and they reflect their forms in matter in crystal structures, which are the only forms of zero curvature in Nature.

2. All motion in all the electric universe is curved. The curved universe of matured form is based upon the sphere. The sphere is a compressed cube. The sphere is a series of true circles no matter where it may be divided into sections by cutting through it in one plane, anywhere.

3. All motion is equally compressed in respect to gravity, therefore, all motion is in true gravity centered circles, which multiply into spheres, and again divide into true circles.

4. The Source of energy which creates true circles, is at their very center. The Source of all energy is the Creator. All motion of every nature, whether of thought or action, spins in true circular control around the omnipotent Creator of that motion, and in planes of 90 degrees from a shaft of any extension of motion.

5. No man, nor creating thing, can in the least way, become off-center from God, or can his orbit around God be aught, at any time, than a true circle, even though its seeming orbit is an ellipse. We think of orbits of planets as being elliptical. And so they seem, but such illusions in Nature are multitudinous, and must be overcome by comprehension. See Fig. 46

6. Man's body is compressed motion, which encircles man's invisible Mind-Source in true circles. He can never depart from these true circles. His thinking and his forming body cells encircle them without knowing that he is but manifesting his own immortal Self, or without knowing that the immortal Self which centers him is continually whispering to whatever of Mind-awareness has yet unfolded within him. The noise and turmoil of

outer-sensing drown out those inner whisperings for long aeons, but there comes a day in his unfolding when he begins to hear that Inner Voice. More long ages pass before he knows that the Cosmos of all that IS is speaking to him and that he IS that Cosmos.

7. Gravity controls the moving universe but man has never known, nor even suspected, that his own immortality and gravity are ONE in their centering of his sensed-body. Nor has he ever known that gravity extends to a shaft as his body extends from thought-ring planes to mass, and as equators divide to become pairs and unite to lose their division. These things man must know together with the omnipresence of gravity which but seems to the senses to extend into shafts made up of many points, which are all ONE. When man overcomes this illusion of the senses he, likewise, outgrows the possibility of forming sensed-conclusions such as the nucleal atom, which violates all principle of this electric universe, which Nature is.

8. Man plainly senses the motion of a plane, or boat, or a speeding bullet and thinks of it as a direction of motion. He sees the motion of a wave, or the vibration registered by a harp string, or the cardiogram of a heart pulsation. His senses also interpret these as motion, and directions of motion, not knowing that they are but registering the curvature of gravity control, and that the direction of motion is ninety degrees removed from that direction of gravity control.

9. That which cannot be sensed in invisible space is yet to be dynamically known. The mystery of the crystal, and its cleavages which divides crystals into units, lies within the knowledge of space geometry, which can be scientifically organized into Mind-visibility, as bodies in motion can be scientifically organized into sense-visibility. There is nothing in nature which the senses can feel that the Mind cannot know.

10. The great mystery of ali of the mysteries of matter lies hidden within the inert gases of the wave. The nature and structure of these keys to motion has never been known. It is time that they must be known.

All of these mysteries shall be briefly touched upon herein for the express purpose of giving enough knowledge of the true nature of the atom to prevent it being used to endanger the human race.

V

Not one action of motion by man or Nature—in all the universe—can escape from the absolute centering of its action by the Controller of all action. Anything, or any man who tries to escape from it, or use his "free will power" to do as he chooses, may exercise that privilege, even unto destroying himself by so doing. He will not escape, however, from the perpetual control of his universal energy Source. No matter what he does, to upset the universal balance, he will find that the Creator of Creation always centers him, and his own unbalanced orbit is still on center with his unbalance, and is perpetually in a plane of 90 degrees from the shaft upon which its circlings turn.

Consider a top, for example, which is spinning true upon the planet's axis of gravity. See Fig. 46-A. All motion around that axis is in true circles, and in planes of 90 degrees from that axis. The time comes when the top can no longer spin fast enough to keep its spinnings centered by the earth's axis. As it slows its speed its axis inclines. Every turn of the top still spins true to the now wobbling axis. That wobbling axis is compelled to revolve around its Source of balance, however, and in doing so it inscribes true circles around its gravity Source, and they, likewise, are in planes of 90 degrees from that axis, as illustrated in the diagram A-1 and A-2. Consider the earth's orbit, for another example—see B-1 and B-2. When the earth was where Mercury now is, there was but one common center of gravity, and one equatorial plane for the earth, and for the sun. The earth has now left that position of perfect balance between itself and its source and set up an equator of its own, which is at an angle of 23 degrees from the sun's equator. Its axis has a similar variation to that of its source. It has not escaped God-control, however, for the sun's gravity still controls it by centering a larger circle, which includes its entire wanderings around two centers instead of one. Fig. 46-C illustrates this principle in an off-center flywheel which will turn with uneasy and uncomfortable oscillations if slow enough, but

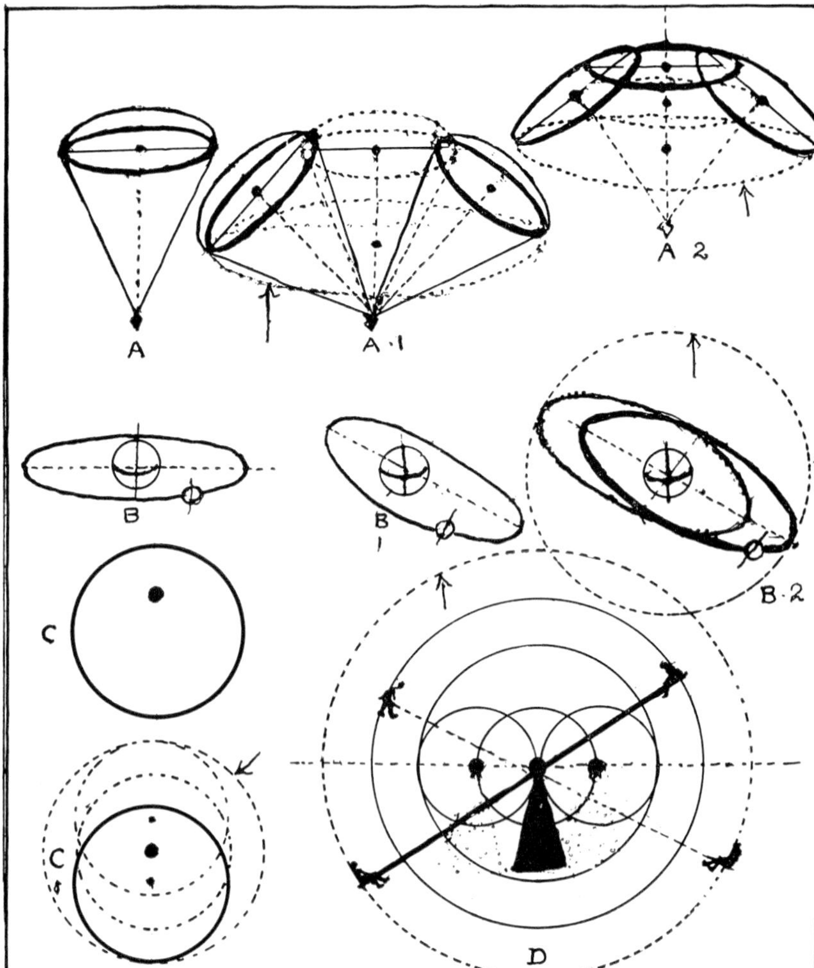

FIG. 46. Every thought and action in this universe is centered by its energy Source, which controls its actions and thoughts from within and balances them from without by two extensions of that center. These extended poles are the ends of gravity shafts around which all masses rotate under true circular control. A—represents a spinning top. B—the earth's orbit. C—a wheel with a shaft off-center. D—a seesaw. See text.

would shake itself to pieces if turned very fast. This example is a good symbol of our present civilization, which is ever turning faster in its eccentric orbit, and ever approaching its dissolution.

Fig. 46-D exemplifies the normal balance of equal interchange in sequential transactions. Every giving is balanced by an equal regiving. As long as that interchange continues its product is GOOD and will continue. If one of the two children upsets that balance by unequalizing the giving and regiving to manifest Nature's love principle, self-made disaster is certain to follow. The disasters of man's civilization are all self-made. Their orbits are ONE with their Source.

Man must some day learn that every cell of his body must be in tune with the universal rhythm in order to keep his body in balance, and every thought and action, which controls his body, should be within as close a circular orbit as possible to reach within to the divine Source of all things where omnipotence and omniscience awaits all men. In this way only, and through such knowledge only, can man know the meaning of that Silent Voice, which has tried to reach through his senses for ages. A continued effort to thus decentrate from concentrated movement to universal stillness, gradually unfolds one's spiritual and intellectual nature. Decentration invites meditation and meditation opens wide the doors for inspiration.

The geniuses of the world are those who have learned to talk with God by transcending their senses and becoming Mind-Beings. In this still very early stage of the unfolding of the man idea he is very conspicuously body-conscious and but faintly Mind-conscious. God's intent for man is that he should manifest Mind—not body. The signs of the times tell us, that more and more Mind-conscious humans are coming into their fruition stage, but mass-man is still body-and-sensation-conscious. That is why such chaos is spreading over the face of the earth. Culture is dying because sensation shuns culture. A Wagner or Mozart is worth billions, and should be nourished. The Wagners and Mozarts of today can be found trying to survive by playing in night clubs, or selling vacuum cleaners. We have no pennies to spare for the arts of peace while so many billions of dollars must be spent upon the arts of war, and in extolling those who kill. Man has chosen to upset his own balance by seemingly violating the inviolate. In the battle of civilization for body-supremacy over Mind, it may well be that the moral, mental and spiritual decline of the last fifty years will continue into another degrading Dark Age. Time is naught, however, in Nature, and man will again arise through the efforts of the few among the many who are now becoming more and more Mind-conscious.

Scientifically, what is happening to mankind at the present time is that he has chosen to distort his balanced circular orbit around his Creator. His free will, supported by his ignorance of his unavoidable close relationship to God, and his unawareness of his dependence on God, has influenced him to choose transient body-sensations, body-comforts and body-values instead of eternal Mind-values. It is not possible for man to continue to distort true circular orbits of man around God, by choosing eccentric orbits or by unbalanced interchange in his human relations. No man, or nation, is powerful enough for that. Fear is the product of such unbalance. *Neither man, nor nation, can survive fear. The universe is founded upon love and man must some day become aware of the meaning of love as related to human survival.* Our present civilization is hurting itself by building its own agonies and ills. It has become physically scientific instead of spiritually scientific. It has no living philosophy to overcome the philosophy of death which is now threatening the world. War has taken forty million lives in forty years, but radioactivity can very easily take four hundred million lives in four hours.

Twenty years ago there was less than three pounds of radium above this earth, yet people are dying today from slow accumulation of that little over the years. Today we are creating 8,000 tons of uranium salts each year which is ever so much more deadly. Not only that, we have 600,000 tons of uranium ore above ground. Who today can tell what the price will be that we must pay for that in human lives and deformed births? Who today can tell the price we shall be forced to pay in future years when the upper atmosphere sends us its bill, and the concrete containers, which have been thrown into the sea, become radioactive Frankensteins? We cannot know for twenty or more years what that price will be, but if the intervening twenty years multiplies the amount of killer metals, which will be above ground in the ratio of its present increase, it is quite possible that the doom of the human race is then inevitable.

All of the above has been written to show that man cannot transcend Nature, or violate its tonal rhythms, by attempting to travel outside of the orbit ordained for him. By so doing he has extended his orbit in the direction of death by expanding his circle to take him farther away from God. Fig. 46 will clarify that for you. Note the orbit of the earth. Its potential was maximum when its oribt was a true circle around its source. An ellipitical orbit, and a straying from its plane of balance with the sun, not only widened its circle of control, but unbalanced its relations with its source by dividing its power by perihelion and aphelion positions in its orbit. The orbits of all the planets plainly mark out the road of death, which all dying things must follow. Can it be that man prefers to choose such a path, or is it because he just does not know?

How long must it be before mankind will know his universe and his Creator sufficiently well to live in love instead of in fear? *The answer to that is dependent upon the length of time it will still take for man to know his universe, and to know God who centers it and him.* As long as man chooses to think, or act, out of balance and *close control* with the Magnetic Light which centers his every cell as well as his Soul-Identity, he but weakens himself by distancing himself from his Source of power. In so doing, however, he does not escape from its absolute control and watchful care over him, no matter what he does to hurt himself.

All of the examples given in Fig. 46 illustrate the principle of balanced control of all creating things by their Creator. It also illustrates the principle of multiplied power which man gives to himself by ever drawing the circles of his body in closer circles of balanced relationship with his Maker in order to become a Mind-Being primarily, and a body, secondarily. The secret of power, therefore, for any man, does not lie in his ability to *create* motion. It lies in his ability to knowingly *control* it. *A little knowledge is more powerful than a big cyclotron.*

CHAPTER VIII

The Oneness of Gravity and Magnetism

I

WE HAVE NOW ARRIVED at a point where the relationship of Gravity and Magnetism can be more easily comprehended. They both belong to the zero universe as shown in Fig. 6. They are both one, but each have a different connotation in common usage which require two words to define their separate meanings, just as an armchair and a rocking chair are both of them chairs, but require two words to differentiate them. *Gravity really means a point, or shaft, that can be located in the invisible universe, where Mind desires to concentrate thought, and thought-power.* You cannot see gravity but you can locate it at the very center of every creating thought-ring, which constitutes what we call bodies of matter. You cannot *see* Mind either, but you can locate Mind also, for Mind is also that center which balances, controls, surveys and motivates its electric thought-rings in their effort to simulate the Idea existent within that stillness, by fast or slow motion.

There is not *more* gravity or *less* gravity anywhere. The seeming increase of the power of gravity *is electric potential,* which means faster motion in smaller circles. Gravity shafts, and centers of gravity are the omnipotent zero of the Mind Universe. It is the same everywhere. It is *all-powerful* everywhere. The nearer that motion can come to it the greater the power which can be drawn from it. *In man that power is desire. Desire for closeness to God gives one that power in the measure of man's desire. The measure of power which man takes from it is the measure which is dependent upon his own desire for omnipotence, and his willingness to multiply his thought-power and action into electric potential by action. In so doing he decentrates to the Source of where gravity multiplication begins in order to conceive Idea and gain knowledge. He then concentrates to manifest God's omnipotence in him.*

Gravity and the Magetic Light are ONE, but the connotation of gravity differentiates it from the word Mind, or Light, just as the word Soul differentiates it from the word God. They are the same, however, for God is the Universal Soul while Soul in man mean's Identity, or Being, as a unit of the Universal Being. The moment that electricity divides the One changeless condition into pairs it is necessary to balance and control those pairs. Gravity is the Magnetic control and balancer, although its power never multiplies or divides. *Where gravity is stillness is. An area of stillness always surrounds gravity shafts.* These areas are the holes which center the rings of electric potential which continually multiply compression, or divide it by expanding, in accord with the desires of Nature, or man, to manifest that all-power which is within, and omnipresent in all things.

The one most important thing to burn into your consciousness is to realize that wherever motion is it is centered by stillness, and that stillness is its CAUSE. The universe is composed of electric thought-rings of motion. Each thought-ring is centered and controlled by the Creator of that thought-ring. God creates His universe that way. *You create your universe that way.* It is God who sits there. It is the Magnetic Light of all-knowing, all-powerful Mind which sits there. We call that centering point *gravity.* The reason why we call it gravity is because there are adjacent thought-rings which have united together so closely that we think of them as *mass,* instead of rings. Each added ring of the mass has its Mind-center in its own plane. In a mass, therefore, there are seemingly so many Mind-centers of stillness that they constitute a shaft. The word *gravity* arises from this fact. *Gravity is a shaft of Mind-controlling stillness. In that sense, gravity and God are one.* Now you have the whole story except for one thought to complete it. Every point of that gravity point is the same point, for God's zero universe is omnipresent.

It may take you many years to fully grasp the above, but you can never comprehend it by reasoning with your brain as a sense basis, for in that process is the motion of thinking. You must learn to "be still and *know*." The inner thinking of deep meditation will take you into the zero world of *reality* where illusion does not exist. The more you are able to find that center where the Light of Creation dwells, the more you will be able to discover the Light of your Self, and that is the greatest miracle that can happen to anyone. *When you find your Self you also find God in His kingdom within you.* When you have thus acquired full awareness of your own omniscience, omnipotence and omnipresence you can then knowingly, and authoritatively, say: "I and my Father are ONE."

Man's unfolding power increases in the measure in which God-awareness in him increases. The space which we cannot give here to explain our meaning is fully compensated, however, by the explanation in GOD WILL WORK WITH YOU BUT NOT FOR YOU. The more you study that book the more you will feel your closeness with God and dependence upon Him. As your power increases through that greater comprehension your command over motion, and over your own destiny, increases in proportion. *The main thing for you to fix into your consciousness is the fact that this is a Mind and motion universe, and that Mind centers motion.* In the Mind is all-knowing and all-power. In the motion is the image form of Mind-knowing which manifests Mind-power without being that power. In that sense it would be more exactly definitive to term it a Mind and Mind-thought universe—for all electric motion is but the record in motion of Mind-thinking. It would be even more simple to think of our universe as a Cosmic cinema, for it parallels the cinema of man in every respect.

II

It is necessary to correct the popular conception regarding what is meant by the term *magnetism,* which is now popularly regarded as a physical force, closely related to electricity. We hear much about magnetic fields, electro-magnetism, magnets, magnetic lines of force and such phenomena. We hear about the power of magnetism to pick up tacks, needles and small iron objects with small magnets, and the power of lifting tons by large magnets. We often hear of the universe being referred to as an electro-magnetic universe. It could properly be so termed but only in the sense which we would apply to a Mind and matter universe, or the term Creator and Creation.

We also hear it said that the earth is a huge magnet, such as the bar magnet made within an electric coil. It is not like a bar magnet, however. The middle of a bar magnet is "dead." The center of the earth is alive. Earths and suns are the result of collisions which occur between polarized pairs where two bar magnets come together as you will see in figures 11, 12, and 13. These polarized pieces of steel should not be called magnets, however. Like the earth they are electrically created. They are electrical effects controlled by Magnetic stillness, which means Mind-stillness. It is popularly believed, however, that magnetism comes from matter which is endowed with magnetic qualities. It has ever been claimed that all of such material must be in the earth's crust not more than fifty miles deep, and that influence creates a magnetic field for earths and suns. It, also, presumably created magnetic fields and lines of force, which bend around from equators to poles. There are many such theories, all of which are groundless, for they are all of them perfectly explainable electric effects which we will enlarge upon in Chapter IX.

Much confusion has been set up in world laboratories because of this concept of magnetism. The physicist deflects electrons from their paths in vacuum tubes by holding a magnet near their path, and sincerely believes that the power of magnetism does that instead of electric polarity. He does not realize that the polarized condition in his "magnet" is an electrically divided equilibrium which has created two strong electric vortices, which his electrons are being pulled into just as particles floating in your bath tub are pulled into its spirally turning outflow.

Hundreds of other misinterpretations of what electricity is doing fills all research laboratories and not only slows their progress but gives them a tremendous amount of unnecessary work in experiment after hopeless experiment, at tremendous cost to find out things which are quite simply predictable to one who knows CAUSE and the WHY of things.

Before passing this subject it may be well to give one more example of misinterpretation where electric effects are attributed to magnetism. In the astronomical laboratory they use an instrument called a spectroscope which tells what the stars are composed of in lines of colored light. Michael Farady "discovered" a relation between "magnetism" and light. Later a scientist, named Zeeman, used that principle to detect the "magnetic fields" of stars by applying a magnetic control to the spectroscopic light lines of the stars. As a result he succeeded in splitting one line into two, making it look like a tuning fork. With a stronger, "magnetic field" he again split those two so there were four, which then looked like a table fork. The result is interesting, and valuable, but it is not what it is claimed to be. It is a division of one fixed condition into pairs, and octaves of pairs, *and that is the sole office of electricity*. Why call the *polarizing principle of the electric current a Zeeman effect?*

The word *magnetism* should in no way be used, for it connotes a *physical* attribute. A proper way to connect it to this electric universe of motion would be to term it the Magnetic-electric universe, meaning the spiritual-physical universe, or the Mind-thinking universe, in the sense that the zero Magnetic universe is the Creator, and the motion universe is Creation. Likewise, we should never use the term *matter* in relation to the visible universe, for the word matter connotes substance, and there is no substance in all of the motion universe. We must someday remodel our concepts to the realization that the God Light of all-knowing and all-energy is all there is in this omnipresent universe of Mind. Mind thinks its knowing, however, and electrically records the imaginings of that thinking by two-way divided effects of motion. Motion simulates that which it seems to be but that is all it does. *It creates a motion-sound-picture by the light which is projected from Mind.*

What observers think of as magnetic power is the measure of electric potential which is created around gravity shafts and centers. The multiplication of electric potential around gravity shafts is—in reality—the manner in which Mind-thinking concentrates.

It is not gravity, or Mind, or stillness, which multiplies, however. That which multiplies and divides is electric potential only, and that means fast or slow centripetal or centrifugal motion. The gravity which evidences Mind-concentrative power is changeless. Motion surrounding it alone changes.

III

The electric current begins and ends at cathodes. Cathodes are still points in the zero universe from which the energy of desire for creation is expressed. Chemically, cathodes are the inert gases of the octaves, which are not elements, for they will not mix with them. They are the seed from which the elements spring, and to which they return. From the spectrum standpoint they are white light from which all colors extend when put under electric strain and to which they return when the strain ceases. From the tonal point of view they are the keynotes of the octave from which one can never escape knowledge of their presence in every tonal harmony. From the mathematical point of view they are the zero of the whole octave, which is more fully described in Chapter XI. From the geometric form point of view, which its basis for motion gives it, the inert gas consists of four rings, one within the other.

1. Magnetic poles are "created" by coiling a wire around a bar of steel in such a manner that the compressing rings of the electric current will produce these so-called magnetic poles at the ends of a bar of steel. That is what electricity does. It produces a potential of electric motion around a still center, but the still center is gravity. To be correct *these poles should be called gravity poles,* connected by a gravity shaft. They control the sex-division of Nature. The Magnetic Light is sexless for it is in equilibrium. Its electric division into pairs creates the dual sex condition which we know as male and female. When these two conditions unite they become ONE. Gravity poles are TWO, when divided by the electric current. When they unite they become ONE CENTER OF GRAVITY. We have always called them magnetic poles when they were TWO but when those two were united we have then called them a center of gravity, and that is not consistent.

2. Electricity, therefore, produced the poles by compressing

the holes out of the rings. They did not come there in any other way. Electric motion does not *create* gravity poles. It merely locates existent points and makes them *identifiable.* It makes the invisible become seemingly visible. However, it never occurred to early observers that the division of God's stillness into sexed pairs would have to have a measure of balanced control over such a division. In other words, if the one condition of balanced stillness in the zero universe is divided into two unbalanced conditions, it is then necessary to have two controlling points of stillness around which motion can spin while thus divided.

3. The two poles are, therefore, poles of still Magnetic Light around which the divided electric pairs can manifest their light of motion. The two poles, thus divided, have within them the desire to be united as one, by uniting the two divided pairs of conditions as one. In this manner the two poles which have united, become the common center of gravity of the whole gravity shaft of the mass controlled by it.

4. Magnetic poles of gravity are still centers which balance and control the electric motion which surrounds them. Electric motion can be insulated from all other electric motion, but magnetic polarity, or the gravity center where poles unite as one, cannot be insulated from anything, for they are of the omnipresent stillness upon which the universe of motion is based.

5. Neither polarity nor gravity are "created" by electricity. They merely become points which can be located in the omnipresent stillness. When, therefore, we say that electricity creates the condition of gravity, we mean that electric motion is spinning around a point in the omnipresent vacuum which controls that motion.

6. Polarity begins as sexless unity at the cathode, which is the location for the inert gases in the octaves of elements. From there it is divided into sex-conditioned pairs as they are extended centripetally toward anodes. They again unite as one sexless unity at amplitude wave positions where they collide and become the two hemispheres of an incandescent microscopic, or macrocosmic sun, which corresponds to a "loop of force" in an electric current.

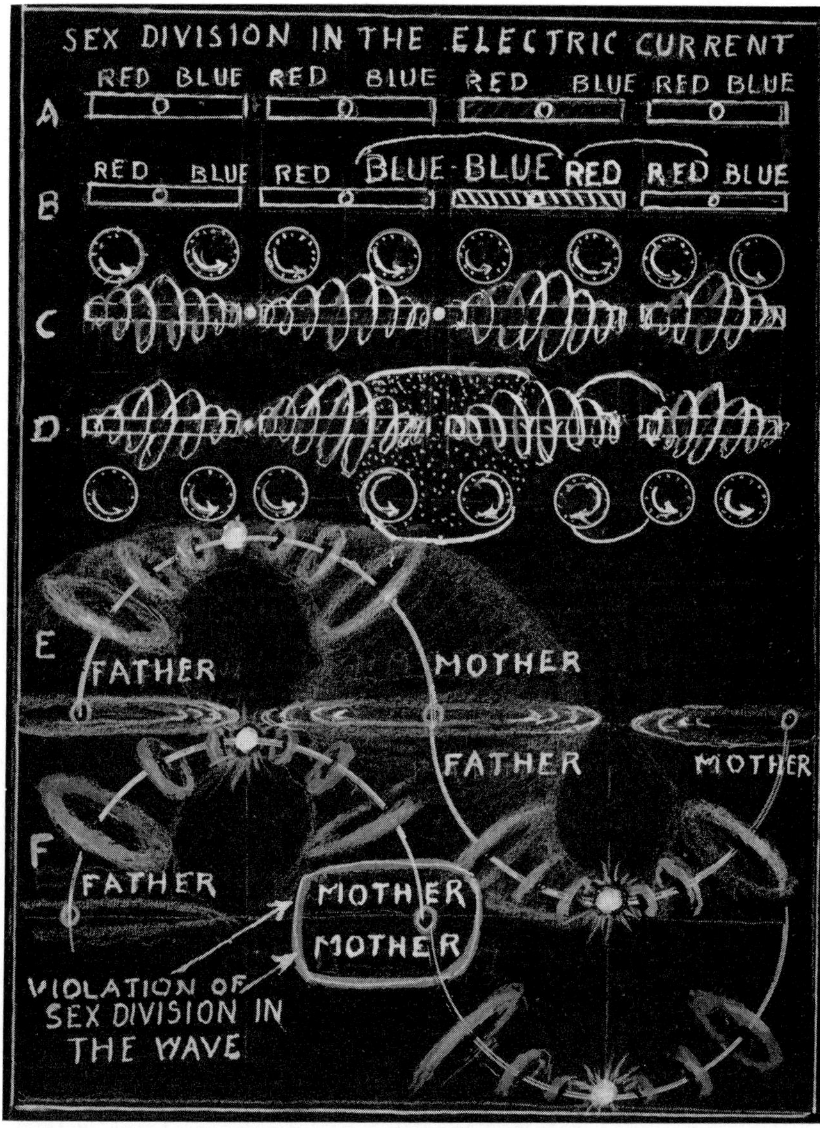

FIG. 47. Nature arranges her sex divisions in unit pairs in such a manner that opposite sexes can unite as one. Nature will not allow man to reverse this order by turning one unit of an electric current, or wave, around to make sex unity impossible.

7. The order of creation, as expressed by the electric wave, electric current, or magnet is as follows: RED—O—BLUE RED—O—BLUE RED—O—BLUE RED—O—BLUE. Fig. 47-A is a series of bar magnets placed in this order. You cannot take one of these bar magnets out of this order and reverse its ends—as shown in B. If you do so you have two males and two females where the normal mates of opposite sexes should be. This would result in the following impossible order RED—O—*BLUE* BLUE—O—*RED* RED—O—BLUE.

8. Electricity spins around all centers in this whole universe in one direction only. C. represents this universal direction of electric spinning around the same series of bar magnets. That one direction is clockwise, or anticlockwise, according to the position of the observer. If you look at the spiral turning from one end, in the direction of the other end, you will see a clockwise direction. If you look at it by reversing your position you see it anticlockwise. The direction of turning does not change because you change. You cannot pick out one half of a cycle, however, and reverse it as indicated in D.

9. The most important new thought in relation to electricity is that every action creates *only one cycle of an electric current. That one cycle is the electric record of that action.* The omnipresence—or Oneness of the universe—repeats that record throughout the universe. The motion of your little finger, or of a dynamite blast, will create one cycle for each action, for electricity is the recorder of Mind-thought, and Mind-thoughts are consecutive.

10. Future laboratory techniques should eliminate from two thirds to nine tenths of its wiring. This will follow a greater knowledge of the geometry and mathematics of space, and especially the tonal nature of octave waves, to make it possible for cathodes to find their anodes without wires, as they do in Nature. The cylindrical solenoids of today should be entirely eliminated, especially where they encase anodes and prevent their expression of power by the way of equators. Vacuum tubes for every purpose should be entirely re-designed to conform with the action-reaction mathematics of the cycle, which accelerates

and decelerates in cube ratios, as potential is multiplied for one half of the cycle, and divided for the other half. The cylindrical coil ignores this law of Nature, even as Einstein ignored it, and makes very clumsy affairs of step-up transformers and a tremendous complexity in electronic machines, which is entirely unnecessary.

It is said that storms in the northern hemisphere are anticlockwise and are clockwise in the southern hemisphere. That is because you are looking down upon it from one end, then you are looking up into it from the other end. *You would see the same effect if you look into the spiral of your bath tub drain, then look up from under it.*

Fig. 47-E illustrates the fact that you cannot pick out one section of a wave, or of an electric current, and reverse that one section. If you do this you will reverse the universal direction of motion, and the universal division of sexed pairs. Nature will not allow this to take place. See 47-F.

To account for this unnatural phenomena The Coulomb Law was adopted, which says that opposites attract, meaning oppositely sexed mates, and likes repel, meaning similarly sexed pairs. This law is invalid, for oppositely sexed mates do not attract. They are forced into collision by the electric action of dividing Oneness into unit pairs. The instant that they do collide they use their utmost endeavors to re-attain the sex-divided condition by recharging their discharged condition. Nature helps in this process by assisting to re-charge with every heartbeat, every breathing cycle, and the food one eats.

Space will not allow further amplification of this process in the nature of electricity, for enough has already been given in the high points of essentials to serve the purpose of this book. It is not well, however, to leave this mystery of how "magnetism" picks up iron nails and "attracts" steel needles, unexplained, for the entire electrical engineering world is paying heavily for lack of this understanding in many ways, the more expensive one being the vast wastage caused by building improper coils, solenoids, armatures, step-up, step-down transformers and electronic tubes.

The element iron—like cobalt—is formed almost at the very amplitude of the wave. Its position is almost at the collision point where mates find unity in each other. Also, iron is on the red side of the spectrum division, and the red side bores within the blue when they seek unity. Nickel, cobalt and copper are on the blue side of the same octave. Because of this position in the wave, iron and cobalt are so constructed under high compression and high melting points that they "remember" the motion of the electric coil which borned them, even after that coil is removed. The spinning effect still continues within them, and will still continue for many years unless they are heated to a sufficiently high temperature to explode the power of electric potential, which these poles have accumulated. Conversely, cold multiplies that memory which heat destroys. At absolute zero, polarity and conductivity are both more intense. It should be sufficiently convincing that the "magnetism" attributed to this electric effect could not be an existent force, for if it were it could not be destroyed by cold. Factually, *cold multiplies electric potential,* and *heat divides it.* This fact of Nature should bring to an end this unnatural concept which attributes electrical effects to something other than electricity.

Copper and nickel occupy the same relation on the blue side of their octave that iron and cobalt occupy on the red side, yet neither of them are able to retain the memory of the electric coil which borned them. The reason for that is because the blue side seeks the outside of mass and, therefore, has lower densities, with consequent less power to retain a "memory" of the motion.

Regarding the compass needle which seeks the point of north at one end and the point of south at the other end, the principle is the same. Every compass needle is a miniature bar magnet. If you place a lot of iron filings on paper, in the familiar way shown in all text books, and shake them over a magnet, those findings will form curved lines. These curved lines are called magnetic lines of force on that paper, and also in the so-called magnetic field of a planet. They are not magnetic lines of force. They are merely the curved planes of opposing pressures which electricity is causing in its efforts to compress. The vortex of gravity which is in the compass needle, will follow these pressures. Every effect of motion in this universe is an electrical effect caused by an electrical force acting under the control of the invisible Magnetic universe. The invisible universe entirely dominates and controls the visible universe, but *motion* is entirely electric.

CONCLUSION

We have very briefly given the essentials of the construction of matter and the true nature of electricity, gravitation, radiation, radioactivity and magnetism. Even though the entire process is as simple as we stated it in one paragraph at the beginning of Part II, its aspects are many, for its effects and its divisions are many. We have not departed from the pattern laid down in that paragraph during this brief narrative, nor do we depart from it in the 48 lessons of our Study Course. Beyond that is still the need for further volumes, but this simple principle cannot be exceeded, no matter how many volumes are needed for greater amplification of essentials.

There are still many essentials which we have not touched upon, such as space geometry and mathematics which determine the pressures of the chemical elements, in order to master and control the principle of transmutation, which is now right at our doors, and the control of drought. Even in all of these it is impossible to exceed the one principle laid down in that paragraph in any effect of motion in this entire universe.

CHAPTER IX

The Mind Nucleus of the Atom

I

TWENTY CENTURIES AGO the consummate Illuminate of all time tried to tell the early man of His day that God centered His universe and man. He freely told man that the kingdom of heaven was within, and that the Father of man dwelleth within man. His words had no meaning in those days of little knowing. The people of His day demanded an objective God, a ruler over men, a personal God outside of themselves who had human emotions. That was the pagan concept of pagan intellects. Very slowly, down through the ages, the Mind-God of a Mind-created universe, began to dissolve the pagan concept of a God outside of His Creation, to a Mind which is omnipresent within it. Today millions of people have entirely discarded the pagan concept and have accepted the Mind concept without comprehending its full meaning, but man in the mass, throughout the world, still holds the pagan concept. Mary Baker Eddy, the Illuminate of the last century, advanced the human race intellectually in that respect more than any other being who has ever lived since the days of the Nazarene. Civilization progressed dynamically and spiritually as man's concept of God thus changed. With even this progress the words: "Seek ye the kingdom of heaven within you," have no dynamic meaning.

The human race can never become ascendant until it is as fully aware of God Presence within man and all things, as it is aware of objective things. The brotherhood of man Idea can never come into its full meaning, and practice, till that new day of enlightenment.

In the month of May, 1921, other words of the same meaning as those uttered twenty centuries ago, were written down as another Message to man to tell man where to find God. These are the words of that Message:

> *"For behold: I am within all things centering them. And I am without all things controlling them.*
>
> *"I center My universe as My knowing. My universe encircles Me as My thinking."*

These words—which are a fragment quoted from The Divine Iliad Message—are fraught with Cosmic meaning. They were written to tell to a newly dawning Age that the nucleus of the atom is the still Magnetic Light of God, the Creator of the atom, and that the atom is the electrically divided pair of moving lights, which manifest His thinking.

God centers His universe. God holds every atom of it together to manifest Him by its purposefulness. God gives of Himself to all His universe in an eternity of endless regiving. God's universe regives of itself to God in an eternity of endless giving. That which God gives is Love. That which is regiven is love. That is the divine story of Creation. *It is a story of cause and effect in the giving and regiving of love.* It is the One story of God's knowing, expressed by His thinking, illumined by the Light of His imagining. It is told in the language of Light projected from His existent stillness into existent stillness to simulate motion where naught is but stillness—to simulate dimension where dimension is naught—and naught else is.

There *is* naught but God—and God's knowing—and His thinking—and His imagining—in all this seeming universe of Mind, which alone is. These are the things which long ages of time, and experience fraught with the agony which is a part of ignorance, have tried to whisper to man from within his inner Consciousness. The silent Voice of God has been ever active in this respect. Intuition and inspiration have also been ever active in this respect. God's Messengers to man, the geniuses, the Cosmic Illuminates and the rare mystics have ever been active in this respect, even though man crucifies them for their service to man.

II

The time has come when unfolding Intelligence in man should tell him that the divine spark of inspiration, and the Silent Voice which speaks to him from within, is the Magnetic Light of Mind and the Source of his energy.

We have been saying all through this book that it is time that all men should know our universe. *We shall now add to that by saying that all men should know God and His universe as One with themselves*.

God centers all units of His Creation. It is God's energy which created His universe. God is the Source of energy which causes all motion. God's Mind is *your* Mind. He extends the energy of His Mind to every atom which moves around His center of stillness where He takes His omnipresent stand in every thought-ring and cell. Whatever knowledge you ever acquire comes to you from your Mind *within*. Your senses cannot acquire knowledge. *They can but be informed of effects of motion. And they can be mightily deceived by what they see.* Knowledge can come only through Mind. Mind-awareness in the human race has but begun. With Mind-awareness comes God-Consciousness.

God-Consciousness and cosmic awareness of the Light of the divine Presence within every man is the next step in the spiritual nature of man. Thousands—and then more thousands—are beginning to comprehend that Inner Voice of one's own Mind coming from within each man. The reason for that is because the human race is still in its intellectual infancy. The Dawn of Mind-Consciousness is too recent for all men to THINK with their Mind. Man-in-the-mass still senses with his body. His desires are still dominated by his senses. His concepts are still sense-based. He hurts himself with his own acts and calls them evil. He conceives a personal God with vengeful human emotions. Evil was conceived in the senses of man. It has no existence in the Light of Mind.

All that Jesus told the world about the unity of God and man, and the location of the Mind-kingdom *within* every man, has no meaning except for the very few illumined ones and the very many who now so deeply desire that illumining. Yes—God-Consciousness is slowly infiltrating the human race as man knows more and more how to THINK with his Mind instead of SENSING with his body. We, who know God in us, are desirous of illumining the path for you to find that Light of inner knowing—and when you do find it, you will then be able to say—with deep knowing—I and my Father are ONE—*and you will know all things.*

III

The nucleus of every atom is Mind-energy. The electrical power of motion, which spins spirally in one direction throughout all the universe, records Mind-knowing in Mind-centered rings. Electric rings are atoms and united pairs of rings are atomic systems and cells. Compressing atoms multiply to express life to its maturity in cells and systems. Cells and systems then divide again into expanding atom rings to rest in their eternal cathode Selves, *which man mistakes for death.*

All motion is Mind-thinking and rest from Mind-thinking. *Thinking is cyclic but the energy Source of thinking is eternal.* All of the energy of all Creation is in the omnipresent vacuum of the zero universe. *The error of man's observation of universal FACT lies in his beliefs that energy moves.* The omnipresent vacuum never moves. It is the fulcrum from which motion draws its energy to move, but the fulcrum never moves. Every ultra-microscopic point in this Cosmic vacuum tube, which the universe is, is a fulcrum from which Mind-desire is extended to express the Idea of Mind. Mind-desire is the sole energy of the universe. Motion is but the lever which expresses the energy extended to it from the fulcrum. As God's Mind is omnipresent, so, also, are all of the qualities and attributes of Mind, omnipresent. That is what man of the future must know and comprehend. He must know that all-knowledge and all-power exists at every point in all of the universe. When he finally knows that he will know that all-knowledge and all-power are within himself. *Man's greatest lesson of life is to become aware of that all-Presence within the Light of himself.* That is what Cosmic Consciousness means. It means becoming Mind-conscious rather than body-conscious.

Let us illustrate the meaning of the above in this way. If a man looks into a mirror he fully believes that he is looking at himself. That is the belief of man for long ages. He is not looking at *himself,* however. He is but looking at the rings of motion which are spinning around *his eternal Self* for a period of activity between a rest period, in which he is building an image of what he thinks his Self to be. No man can see his Self but he can KNOW his Self. *Seeing is electric sensing. The senses are motion. Motion can sense only motion. They cannot sense the stillness of eternal balance. They can but sense the motion of divided balance.*

When a man stands in perfect balance he cannot sense that condition. The very instant he falls out of balance, be it ever so slight, he is then aware of it. The electric current of motion vibrates within his senses and his senses become electrically aware of it. *He misinterprets the motion of his senses for Mind-thinking.* If a man is perfectly comfortable his senses are not aware of it. If he becomes cold his senses *then* tell him of his unbalanced condition in respect to his environment, and he puts a coat on. His *body* has told him these things, however, and not his *Mind. The measure of unfolding Intelligence in man is the measure of his Mind-knowing in relation to his body-sensing.*

Has man unfolded intellectually to the extent in which he can hear God's Voice inspiring him to become co-creator with Him by interpreting His qualities of love, beauty, rhythm, harmony or balance? Or is he limited to fulfilling the demands which the motion of his body cells are demanding of him? This is the nature of the knowledge which will give the coming race more geniuses and mystics, and great leaders among men for the upbuilding of an enduring civilization of peace and good will on earth. This knowledge can be acquired only through desire for it. Desire can be awakened, and then multiplied, through knowledge. To acquire knowledge look within your own centering Self, for it is there awaiting you. The Silent Voice of your eternal Self forever whispers its omnipotence and omniscience to you. The following diagrams, which accompany the words of this chapter, are for the purpose of giving you that knowledge of *where* your power lies awaiting its manifestation by your body.

IV

We will begin at the beginning and picture for you one thought-ring of motion, which constitutes this entire electric universe. Bear in mind that there is not any other form in Nature than this one alpha-omega form, and combinations of them. Fig. 48 represents a single thought-ring of motion, which surrounds its motionless Mind-center. From this Mind-center all-knowledge and power is extended to the electric thinking which records the Idea of Mind. This universe is, therefore, composed of Mind and thought-motion. Its thought-pulsations multiply to the high potential of maturity and then divide into the rest from which they were born, to be again reborn for another period of motion. This figure represents the Inner Voice which speaks to you inspirationally, if you are sufficiently Cosmic to be aware of it, or to lower animal life instinctively.

Following this diagram are others of a sequence which demonstrate that bodies are created and given life, for a purpose, by increasingly fast thought-power motion around their Energy-Source. After they have fulfilled their purpose they must give back to their Source that which has been given to fulfill the law of love, which demands equal giving and regiving. *It must be noted that Nature never TAKES—it only GIVES.* The following diagrams illustrate this principle fully. Each one of them is the prototype of one cycle of an electric current, or an electric thought-wave. There is no other expression of motion than this in all Nature.

In studying these illustrations fix upon your Mind the one thought that the entire omnipresent vacuum, out of which motion appears, has all-knowledge, all-power and all-presence. What happens anywhere in it happens everywhere in it. Your senses may see it only at one point but your Mind knows that it is omnipresent. The radio and television tell you that. Every schoolboy is conscious of omnipresent actions through his familiarity

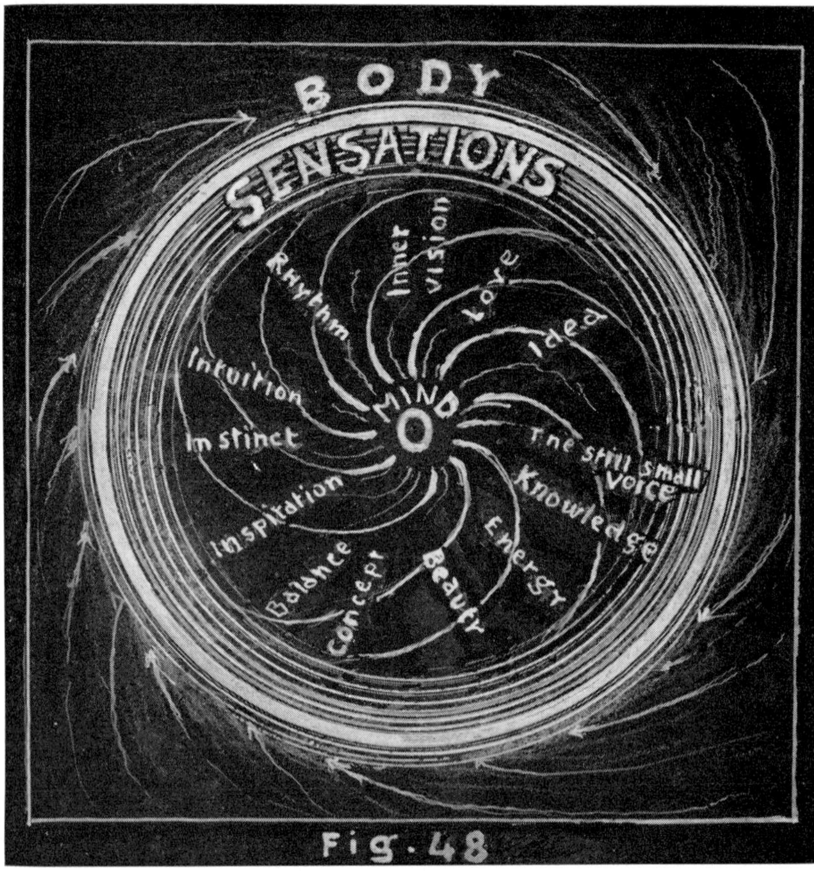

A MIND-CENTERED THOUGHT-RING OF MOTION

Fig. 48 is a section of a body cell, magnified millions of diameters, showing relation of Mind to body. All matter is composed of thought-rings of motion compressed together in spherical cells, atoms, planets and suns. Thought-rings are what we call our senses. They are Mind-centered and Mind-controlled through electrical messages sent to them from within, and are received as instinct, intuition, inspiration or other states of intelligence in accord with the ability of a body to recognize the nature of the message thus received. Man alone has conscious awareness of the divine nature of the messages, but man must be far on his way to spiritual unfolding before he is fully aware that they are directly from the Creator to his Self. These centers in the omnipresent universe of Mind, constitute the "kingdom of heaven," which Jesus bade all men to seek.

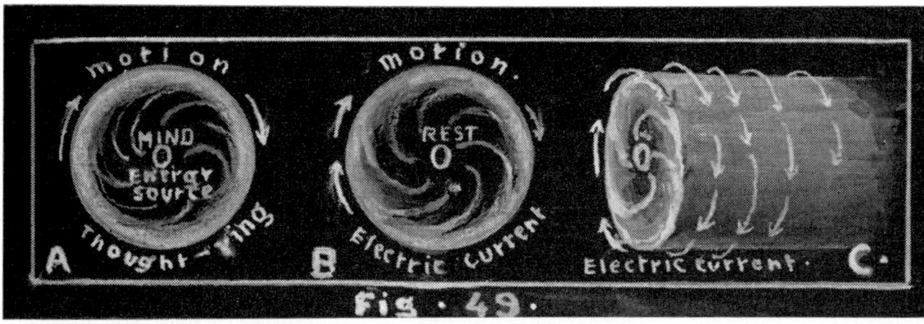

A. Magnified thought-rings moving around Mind-Source.

B. A sliced section of an electric current, illustrating motion around its Source of energy from which it was extended.

C. An electric current sent through a solid cable of this dimension would be a series of rings, like A and B. It would be a very weak current for there would be a very large hole of non-motion within it.

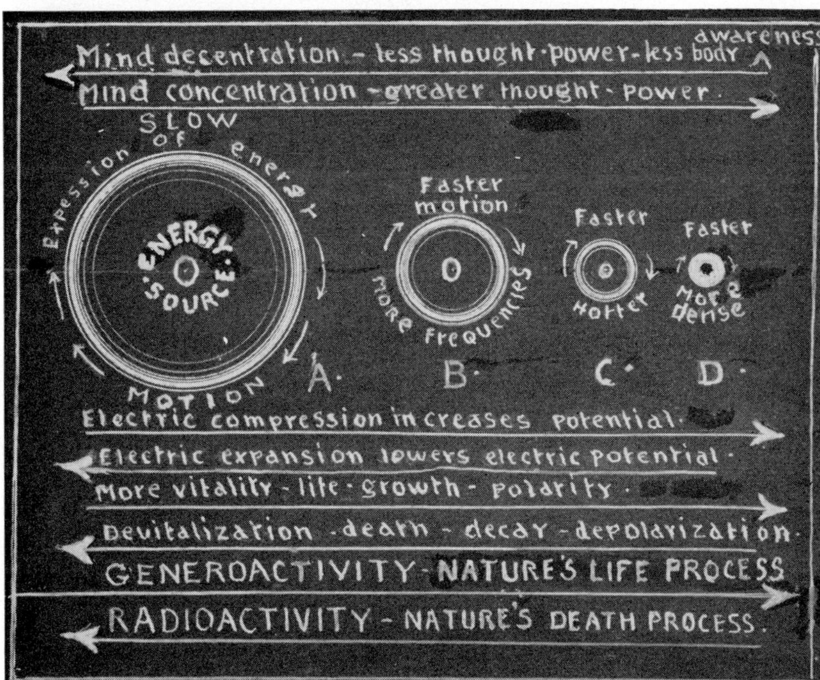

A—represents low potential. The gradual increase in power of expressing the energy extended from the Source, is in the direction B—C— and D. Motion multiplies as thought intensity becomes concentrated, or as electric rings become compressed. Great expression of Mind-energy by electric motion as speed of centripetal force is multiplied. Life itself is possible only through fast motion around its energy Source. Every creating thing in Nature multiplies its life processes in the direction from A to D. It then multiplies its death processes in the direction from D to A.

The polarizing principle. This diagram represents the multiplication of power by the multiplication of motion. Electricity "creates" gravity in this manner.

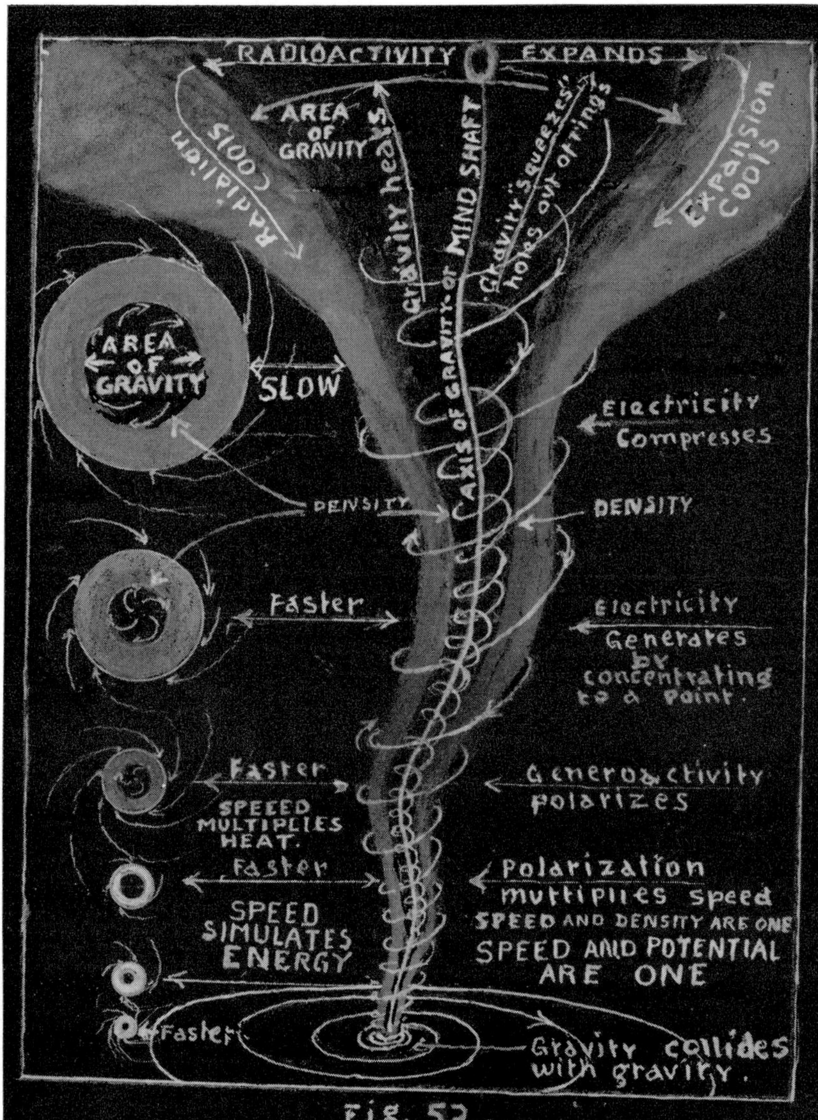

Electric compression principles. Analysis of the relation of stillness to motion, and of energy to the expression of energy.

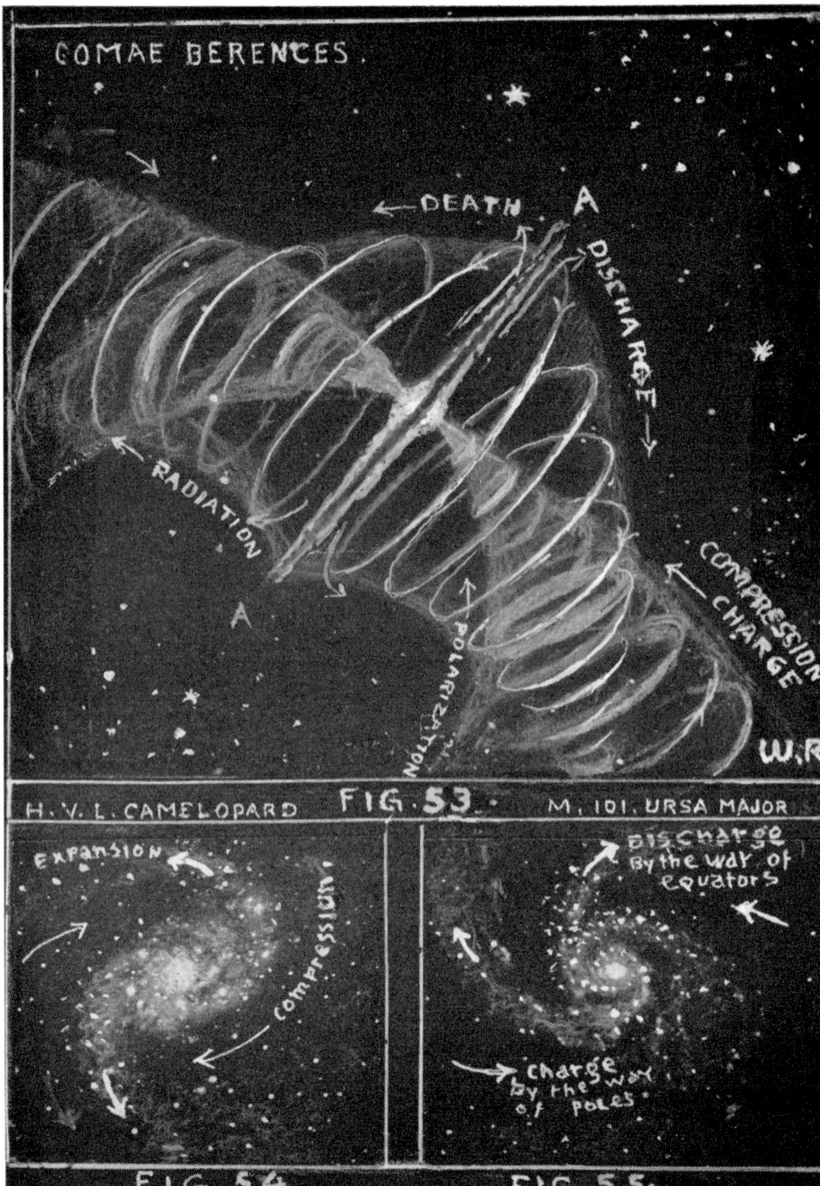

All matter is formed by projecting and compressing light rings in pairs toward each other. The collision of such an impact is a sex union. Only by sex union can bodies be created and repeated. Atomic and stellar systems are formed this way. In an electric current such systems are called loops of force.

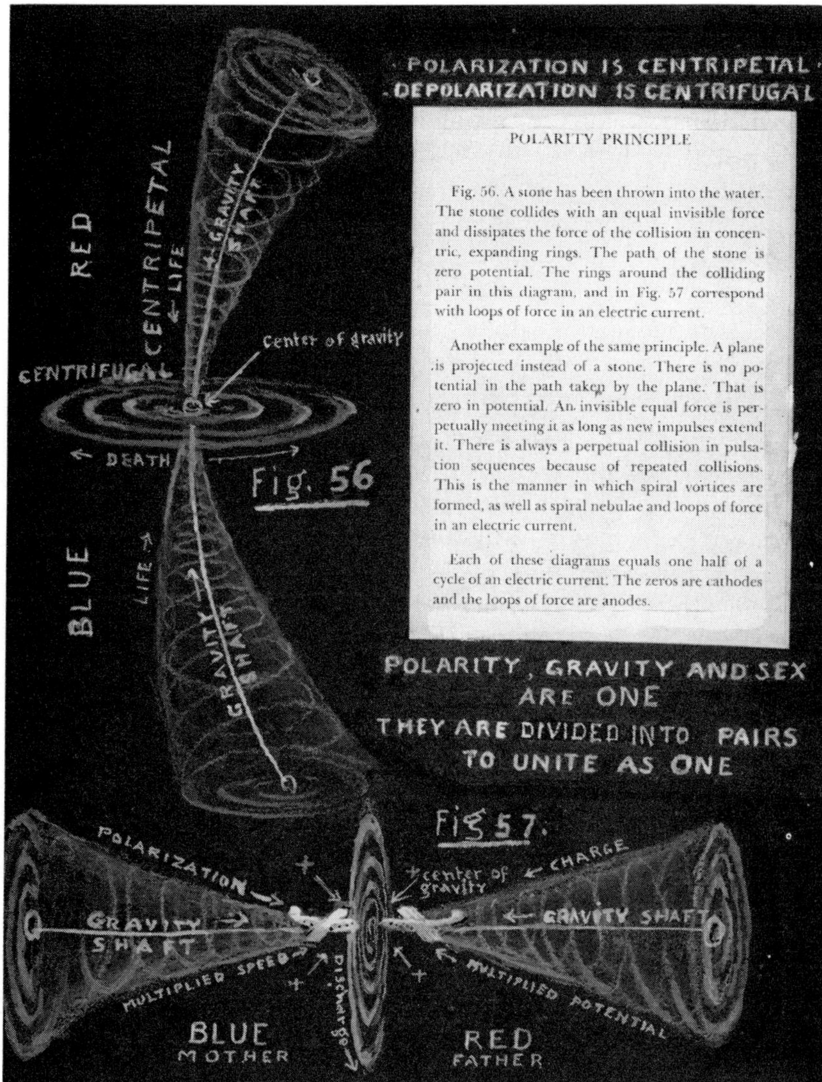

Every action in the universe is created in pairs. Its reaction is a part of it, and is simultaneously created. The union of every action with its reaction *produces an effect which is always the image of the cause of the action.*

with television. That fact he has learned from the radio and television. There is still another greater fact which lies ahead for future generations of Cosmic man to learn. That great fact is that all-knowledge and all-power exists within him and can be acquired by him for the asking in whatever measure he shall desire. To demonstrate our meaning we refer you to the two cyclone drawings figures 51 and 52. In these drawings is a measure of force, expressed by motion. That measure of force is what the cyclone "asked for"—and no more. If it desired to be a bigger and more powerful cyclone it would have more power extended to it for that purpose, for *unlimited*—or OMNIPOTENT—power centers it. That principle applies to all Creation.

A man is what he desires to be. All-knowledge and power are his for the asking. They center him. They are his kingdom of heaven, which is within him. Whatever he desires and asks for, with full awareness of his Oneness with his Creator, he can have by working *with* God to create it. That is why one man knows more than another, or has more power than another. He has learned of his closeness to his Mind-center and knows how to ask for it, and the conditions upon which it will be his. The only way it can be his is to know how to work knowingly with God by becoming one with Him as Co-Creator. The more that one can feel God's presence, moment by moment, the more his Mind-awareness multiplies. That is the kind of mental progress which makes man aware of the unlimited knowledge and power which is omnipresent everywhere, and can be expressed around any point of gravity which is chosen as the center of that expression. This idea is beautifully expressed in The Divine Iliad Message as follows:

> *"Desire ye what ye will, and behold: it standeth before thee. Throughout the aeons it has been thine without thy knowing, e'en though thou hast but just asked for it.*
>
> *"Sit thou not and ask, acting not, for unless thou reach out for thy desire it shall not walk thy way to thee, unaided by thy strong arms.*

"Desire ye what ye will and it shall be thine. All my universe will give it thee in the strength of thy desiring, and in the strength of thy action in reaching out for thy desire.

"Behold I am within all things, centering them. And I am without all things, controlling them. But I am not those things which I center and control.

"I am the center of My universe of Me. Everywhere I am, I am the center of all things, and I am everywhere."

For full comprehension of God's words, as expressed above, one should dispel all idea of the fulfillment of desire by wishful thinking and words of supplication—unfollowed by action. The last paragraph of the above is scientifically demonstrated in Fig. 6. The entirety of this idea is expanded and exemplified in "God Will Work With You But Not For You."

V

Your Mind is universal. It is God's Mind. When you have discovered that fact you become co-creator with God. Until you do discover it, God speaks to you through instinct. When you finally do discover it all-knowledge comes to you from that still small silent Voice which extends inspiration, beauty, intuition, rhythm, and all of the Mind-qualities you did not have until the dawn of Consciousness began to awaken a realization of them in you. Realize then, that your body consists solely of electric thoughts and all electric thoughts are rings of visible electric motion centered by the still Magnetic invisible Light of your Mind.

If you concentrate your Mind-thinking, the thought-rings will become smaller and spin faster. If you relax by decentrating your thinking, the holes will become larger and the rings will spin more slowly. See figures 48, 49, 50, 51, and 52.

If your thinking is unbalanced by emotional disturbances, or your actions reflect your unbalanced thought-decisions, the tensions which result from such unbalance will cause every Mind-center, which controls the balance of those thought-rings, to act as though they were off-center and the spinning of the electric current around that eccentric center will be like a fly-wheel with its shaft eccentrically placed. *When you understand this you will know why your body develops toxins which cause your various ills.* See also Fig. 46 which further demonstrates that you cannot become off-center with your Mind-Source. These diagrams tell you just where your Mind is in relation to every one of the billions of cells which constitute your body. All of them have a central "switchboard" in your brain which sends messages of command to them.

Remember always that a thought-ring is a sliced section of a cell—small and large—of your body. No matter where you slice through your body, or any body in Nature, you will find that each section is a ring with a hole in it. The hole is where invisible Mind is located. The ring is composed of electrically sensed motion that obeys the command of Mind, which centers it. It can but obey for it has no intelligence or energy of its own. Consider such a thought-ring-section in the cell of an ant. The ant cannot think for itself. The Mind which created it reaches out with electrical messages, which we call instinct. All animal and vegetable life is controlled that way by their Creator. They have a modicum of free will in accordance with the amount of intelligence which some forms of animal life attain, but vegetable and mineral life have no such power. They are entirely Mind-controlled through their senses.

Every particle of matter, in combination as mass, or single as a unit, is purposeful, its movements are Mind-directed. Cells composed of oxygen, hydrogen and carbon in combination with each other, are performing marvelous actions in your body. They are manufacturing adrenalin and dozens of other fluids by Mind-control. They are knitting bones and flesh together in accordance with marvelous patterns. They are digesting your food and transmuting it to blood. They group themselves into marvelous machines which pump, knit, weave and perform many skills with many techniques. What man-built machines do you know of that can build themselves, then operate themselves without Mind-motivation and Mind-control? You do not know of any, for that is an impossibility of Nature. It is equally impossible for the cells of your body to build and operate themselves. Each one is a purposeful mechanism. It has to be "created" by Mind-power just as you would have to create a typewriter to fulfill its purpose that way.

That is why you must know where the Intelligence and energy is which operates them. Look at Fig. 48 with this idea in mind. Note in it that all that God is can be extended to the motion which manifests God. All knowledge, energy, inspiration, idea, beauty, rhythm, intuition and other Mind-qualities are within every thought-ring of motion in all Creation. Man is the only unit of Creation which has yet arrived at that stage, and only a very small percentage of the human race has risen to that high stage of unfolding in which man has become *aware* of his centering Intelligence. The very large percentage of the human race thinks electrically with its outer-senses, instead of inwardly toward the inner Mind. Man-in-the-mass is aware of *senses* only. Such a word as *inspiration* has only a superficial meaning for him.

Sense-thinking is purely electric. All sensations of the body are purely electric. One who senses electrically can acquire only what information of sensed EFFECTS that his senses reflect to him from objects of his observation. He is limited in his conclusions and decisions to what his senses tell him of bodies in motion. He can never acquire knowledge of CAUSE that way. He can acquire *knowledge* only by thinking inwardly and thus talk to God direct.

VI

THE RELATION OF THE ELECTRIC CURRENT TO MIND-THINKING

We have diagrammed the relation, and location of Mind to matter in Fig. 48. That diagram shows how messages are sent out to creating bodies electrically for long ages before dawning Intelligence makes bodies become aware of the meaning of those messages, or that they are from the divine Source of all Creation which is "within" all things.

We will now parallel this diagram with another to show that

Mind-messages, which are sent to creating bodies from their Magnetic Source of Universal Intelligence, are identical with the nature of the electric current. Before entering into this comparison, which Fig. 49 portrays, we wish to inject a new thought regarding the process God makes use of in the construction of His electrical-body-universe. Instead of using text-book language we will use simple homely phrases to emphasize our meaning: God creates matter by projecting very big electric rings of visible light around still points of His invisible Magnetic Light. By so doing He creates big black, cold holes in His bitterly cold space. In order to create bodies of solid matter in those cold holes He must heat the holes to incandescence, and then freeze the incandescence by surrounding it with His universal, basic cold, to imprison it until it has fulfilled His purpose. He then uses His basic cold to compress series of four pairs of rings into spheres by sqeezing the cold black holes out and letting the four pairs of compressed ring-spheres of light in. This constitutes His entire generative, or polarizing principle of Creation, for the only things created are heat and motion.

All effects in all Creation are the product of heat and motion. Living things are the *product* of heat and motion. But living things must "die". To "die" they must regive their heat back to the cold which created it, and they must regive their motion back to the stillness from which it was extended. God's way of doing this is just the reverse of His generative method. He gradually lets the cold, black holes return within the hot spheres by projecting rings, in series of four, from the equators of spheres until they entirely disappear into their basic, changeless cold. *All matter thus begins to appear as rings of light around black holes, and finally disappears as rings of light around black holes.*

You can plainly see, by this homely description, that there is no process of Nature which allows for material nuclei in its atoms. The nucleus for all atomic structure is the Magnetic Light of its Creator's Mind. Likewise, the energy for the construction of atoms is in that centering Magnetic Light. It is the invisible Light which centers the visible light. It is the energy of God's Omnipotence which centers its simulation of energy by motion. *It is the Idea of God's knowing, which is manifested by moving forms which image God's Idea.* It is the beauty, the rhythm, the balance and ecstasy of God's thinking of His One Idea of the giving and regiving of His love to His Creation for manifesting His love in all creating things.

CONCLUSION

You should plainly see by this time that the intent of Creation is to dramatize the Idea of Love, by expressing it in action and reaction. Love is one idea, the One Idea of Creation. To express Love the Idea must be divided into separated halves. To fulfill Love the two halves must be united as one. Halves of one must be equal, therefore, their actions and reactions must be equal and simultaneously created. The only action which can express Love is the action of giving. The only equal and opposite of giving is regiving. *The One is divided—the two divisions simultaneously exist.* If the action of giving extends from one it equally and simultaneously extends from the other. If equal giving and regiving of each half are simultaneously expressed the product of that fulfillment is Love.

Can you not see, therefore, that the one great age-long lesson of life is to learn how to balance every transaction of life in such a manner that givings and regivings between pairs of opposites are equal? *That is all there is to learn, for that is all there is in Nature.* Creation consists of multiple electric wave units. In every cyclic unit of Creation the Law of Love is expressed to perfection, as demonstrated in one cycle of the electric current.

Can you not see, also, that the reason why man's world is in such chaos is because man has hardly begun to learn that lesson of the aeons? His givings are few but his takings are many. His empires have been constructed upon his takings, which he is now losing and will continue to lose, until he has balanced them with givings.

Can you not see, therefore, that ONE WORLD of PEACE is far away into the long generations of learning that one simple lesson of rhythmic balanced interchange in every transaction of man, as it is in Nature?

That law of man which reads: "Actions and reactions are equal and opposite," should be re-written as follows to conform to Nature: "Equal and opposite actions and reactions are simultaneously created and sequentially repeated." In exemplification of this law we urge a careful study of figures 53, 54, 55, 56, and 57. Every transaction in Nature is repeated in each of these examples, whether it be the casting of a pebble, or the creation of a sun. In an entire solar system its balanced transactions are so perfectly in keeping with the Law of Love that one can calculate to a split second the exact time and position of any planet a thousand years from now. Conformity with that one law is the only way that mankind can ever find peace, happiness and prosperity in a unified world. How long shall it be before you, yourself, shall find it? In that is the answer to world-unity, for *you* are the seed of the world. *What you become the world becomes from your givings.*

VII

WE EXEMPLIFY THE MIND-NUCLEAL PRINCIPLE

So radically different a conception of atomic structure demands much exemplification to replace the old with the new. For this reason we must cite many examples to demonstrate that every creating, growing, living and dying thing in Nature is just a multiplicity of slow and fast moving light-rings surrounding dark holes of low electric potential, centered by a zero point of absolute stillness and unlimited Mind-potential. The new concept is a challenge to the old which must be met by such convincing proof that the new concept is undebatable.

Let us preface the following examples by these text-book statements of facts known to science:

1. "An electric current sent through a wire is confined to its surface. There is no current at its center." Quoted in essence from Tesla, Faraday, Pupin, Millikan and others.

2. "Living cells show a difference in electrical potential between their interior and the surrounding medium." Dr. Louis M. Katz—University of Chicago.

Figure 49 demonstrates this fact in an electric current. To emulate a section of a living cell we will put a salt solution within a ring of conductive material. By placing two insulated wires, with bared electrodes, close together in the center of that ring and connecting the wires with batteries we emulate the charged condition of the living body, which the cell is a part of.

By turning on the current it will be found that the spot between the two electrodes is of *zero potential,* and that many lines of radiation extend from that center and cause a moving electric current to spin around the zero fulcrum center where the ring is located. Not only that but an electrode placed in contact with the ring, will convey current away from it. This effect of electric current extending from a still center to convey its power to the ring, and other outside conductive matter, is identical with the mental example quoted in Fig. 48.

Anyone who may question this fact may not only prove it by this simple experiment but can also move the two electrodes away from the center of the rim so that their centering zero occupies an eccentric position, as the human heart does in relation to the chest, and as the zero center of the heart itself is eccentrically placed in relation to its mass. A different wave pattern will result but the electric qualities will remain the same. There is always a point of stillness which centers the birth point of any action. Motion spins around that still point, but there is no motion *at* that center.

Consider your body itself, and every part of your body, even to its last microscopic cell. Every part of it is a moving ring in section which spins around a hole, centered by stillness.

Now consider every part of your body, one at a time. Every horizontal section cut through your head is a ring of flesh and bone around a centering hole, where the Magnetic Light of your Mind is located. The space between is filled with a material quite unlike the flesh of your body. That material is the broadcasting and receiving station for conveying messages to every last microscopic part of your body.

Consider your spinal cord. It also is a ring in every tubular section of it, which is not only centered by a hole for itself but is confined in a ring of bone which is your spinal column. Both of these are shafts made up of rings, but it must be remembered that every shaft in the body, as well as every artery and vein tube, is but a continuous extension of rings. Even the center itself is a continuous shaft of zero electric potential surrounded by thought-rings of varying potential. The center itself is omni-Magnetic-potential, for that point is where *you* sit with God, as ONE.

Your heart is an electric thought-ring of motion centered by a hole, where your invisible Mind is located. From that center it sends its complex messages out to the very wonderful mechanism, which your heart is, to command its every part to function true to its purpose.

Your lungs and your chest are rings of thought-recordings which are, likewise, centered by four holes, for the carrying of four different qualities of messages from your Mind to the thought-rings which compose your whole body. We will refer to these again in this chapter.

Now examine your fingers and toes, and your arms and legs. Each of these are centered by zero potential holes around which are rings of bone and flesh of varied potential.

Before completing our brief analysis of Mind-relation to the human body, let us step out into the forest and cut practically any growing flower stem, shrub, sapling or tree. All of them are rings in sections of either tubes or seemingly solid shafts. Even the leaf of the tree is centered by a tube from which countless tubes extend to center the countless millions of thought-rings being created by the Magnetic Light which centers every one of them.

If you cut a section of a new-born oak, you will find that it is tube-centered, but if you cut a section of an old oak, you will find that it has seemingly "squeezed" the hole out to attain density in accordance with the same principle of compression which gives density to a sun. Examine it, however, and you will see a series of rings surrounding its center.

If you now examine the chemical elements, such as oxygen, fluorine, helium, nitrogen, sulphur, phosphorous, sodium or potassium, you will find that their nuclei consists of varying sized holes surrounded by rings of varying sizes and numbers. Helium has four rings with a very large hole within them. Helium divides into four pairs and becomes a solid sphere, which has squeezed the hole outside of it. Fluorine and lithium have one ring with a centering hole smaller than helium. Oxygen and beryllium have two rings with still smaller holes. Nitrogen and boron have three rings with almost no holes. These male and female equal pairs unite in marriage and become two hemispheres, the red and blue halves being divided by an equator.

Chemists and physicists have recognized something of this mathematical orderliness and have given it the name of "valence." They say that oxygen has a valence of two, carbon four, etc., etc., and they balance and vary their mixtures in the ratios of valence by putting two atoms, which have a valence of two, with one which has a valence of four, to equalize their potentials. This they have always done, but their text books give no adequate explanation for doing so.

The real reason is that spinning rings which get closer to their Mind-centers in the Mind-nucleus of every atom, gain more and more power in the ratio of that closeness.

Before returning to the Mind and body relation in man we must make it clear to you that the chemical elements must be considered in a very different way than man is considered. All have a Mind and body relation but minerals and gases constitute the first stage of Creation only. They exist as compressed matter before water is created by combination with two of them, and air by a combination of another two. The next stage is a combination of these elements with water and air to create the vegetable kingdom. The simplicity of the mineral kingdom is then complexed by the addition of water and air and more "conduits" are arranged in these combinations for the placement within its creations for the zero centers of Mind-control. The vegetable kingdom has very little free will choice of decision regarding its welfare or destiny. It has some, however.

Next in order of complex combination is the animal kingdom. This third stage includes both mineral and vegetable matter, in combination with water. The animal body is a complex pattern of variously arranged mineral and vegetable life, so must continually be replenished with all three stages of creation as its various parts "die." For this reason very much more space is given to the body for Mind-control, and for the creation of very different forms of matter of high conductivity to carry the messages concerned in body "manufacture" and body survival. Directional messages, such as instinct, have a very different reflex toward matter than those which may be called automatic reflexes, which are very much more keen in animal than in vegetable life.

The fourth—and last—stage is the human. Man's body is also a combination of all three stages, but Mind-awareness only begins in the human stage. The animal has much intelligence but is not aware of it as a quality within him.

Man, the human, was first controlled by instinct. For millions of years he had no Mind-awareness. He had no creative ability, whatsoever, and not even enough sense of rhythm to desire to beat a drum. The desire of the savage to beat a drum is the first evidence of spiritual unfolding in man.

Man—the human—began at the dawn of Consciousness a few thousands of years ago. He became slightly aware of a creative force outside of himself, then more fully so, through the drum-beat stage to the more extensive recognition of universal rhythm in the four stringed lyre up to a comprehension of the diatonic scale in the 15th century. A sense of beauty, as seen through the eyes, had developed much earlier than that, but the jungle sense of cruelty, brutality and blood-pleasure still remained in man-in-the-mass even while his culture was unfolding, and still remains with a very large percentage of the race. Marauding tribes have become marauding nations and man-killing by man is still done on such a large scale for greed and power only, that the human race has no justification, whatsoever, for thinking its present civilization, as a whole, to be in any other than an early barbarian stage. Many, however, have arrived at a very high intellectual point, and a few have reached almost to its summit. It is through these few that thousands of others are being awakened to that Light of their own Oneness with the Source, and it is through these that the race will finally achieve such intensive God-awareness that the man-killing stage of today will have become impossible.

A more comprehensive and expanded explanation of man's early unfolding can be found in *"God Will Work With You But Not For You."* The above is cited solely for the reason that the new knowledge given in this chapter will give every man a better comprehension of how he can lift himself above the present stage. Having thus explained the reason for this interlude, we will now return to the climax of the knowledge for which this chapter has been written.

VII

MESSAGE CARRYING FLESH AND BONE

It is commonly said that man is made of flesh and bone. That is not the whole story. There is muscular flesh, and there is what is known as neuro-muscular flesh. There is also bone and the marrow of bone. Just plain flesh and bone are the physical working parts of the body, but much that is in the body is created for the express purpose of conveying messages from the Mind WITHIN each cell to its surrounding ring, and the nerve-flesh which conveys messages to the farthest extremities of the body with the speed of light.

The vegetable kingdom is practically all "flesh and bone" but even that has an intricate intercommunicating system which is composed of a very different kind of flesh, a kind which corresponds to the nervous system in animal and man.

The muscular and bone flesh which is used for utility purposes, that which we call meat in chickens and cattle, is entirely different from the flesh of the heart, brain, spinal cord and nervous system, and also of arteries, veins and bone marrow. The blood is also flesh of the body, as other parts of it are, and it comes from the Mind-center of thought-rings which constitute the heart. That too is a flesh which is mostly concerned with its message-carrying purpose, both for body-building and Mind-thinking purposes. That flesh is so sensitive to tensions and strains of unbalanced actions and emotions that its normal thought-ring cells quickly change, or their centers shift to eccentric positions very quickly. The slightest emotional disturbance acts upon them in such a manner that they seem to have an intelligent understanding of such tensions, for they immediately convey that tension to the heart, brain and stomach with a consequent equal upsetting. All message-carrying flesh and bone marrow must have the rhythms of happiness and joyousness in order that their normally balanced electrical state will extend normal balance to the cells that they are continually creating and controlling, as well as informing as to purpose.

If people only realized that a happy body cannot exist as a normal body without a happy state of Mind, there would be no "mental cases," or insanity, stomach ulcers or heart diseases.

One little example of such effects should be enough. Consider the red blood corpuscles as that example. The blood is the basic

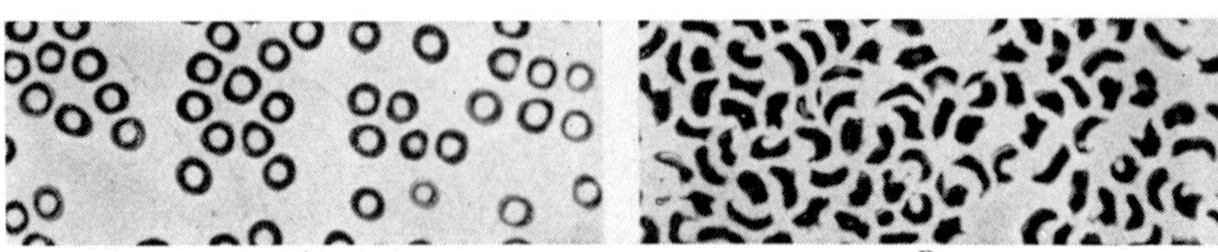

A B

One of the very first evidences of temporary or permanent degeneration of an individual, or a race, is the breaking down of the normal thought-rings of the blood cells from circular to sickle-shaped rings. Any abnormality, or unbalance in man's thinking, and his decisions which cause unbalanced actions, is first shown In blood cells. If this abnormality is not corrected by the individual during his lifetime, his children can inherit the tendency. The whole human race could be destroyed that way if the tendency is not reversed by balancing the unbalanced condition. (These two diagrams are reprints from Scientific American.)

living flesh of the body. Strangely enough—but appropriately—its red blood cells are microscopic rings with holes centering them, such as pictured in example A, shown above. Emotional, or other disturbances, even inherited disturbances, which sufficiently upset the balance of the body metabolism, will break openings in these rings and make them appear to be sickle-formed as shown in example B.

The little white discs which constitute the white corpuscles of the blood appear to have frayed edges, as though they were moth eaten. Anemia and leukemia follow such abnormalities.

The blood is of first importance of all of the elements which compose the body. The nervous system could be entirely paralyzed and the body would still function, but the blood has deep instinctive awareness of its existence and the body, which does not have a happy, rhythmic blood condition, cannot possibly retain its normalcy. The blood immediately feels every mental and physical tension it is subjected to.

Next of importance is the heart flesh itself. There are muscular parts of the heart which perform physical functions only, as other flesh does, but a great part of the heart is composed of an independently living kind of "flowing" flesh which is not dependent upon the whole body, as all other parts are. The blood is the most potent of all of the Mind-message carriers of the body.

If you cut any part of your body out and preserve it from decay in a salt solution it can be thus *preserved* for a long time, *even though it is as dead as the flesh in your deep freeze is dead.* Not so, however, with that part of your heart which is made up of what the text-books call specific, or autonomous neuromuscular flesh. If you put that in a proper electric conducting medium, such as a salt solution, it will *live* a very long time—indefinitely perhaps—if the solution remains properly conductive. It does not die as other flesh dies so long as it is held in a conductive medium. The blood really has no intelligence but it is so close to the Mind as a carrier of power and creative messages that it seems to have intelligence of its own.

There are other parts of the heart which are, likewise, intended for the mental and physical state of the body which are utterly absent in all vegetable life, and less conspicuously developed in lower animal life. Such parts are the sinus node, the right and left bundle branch, and the intricate intertwining nerves, veins and muscles of the lower heart.

More important still is the fact that the blood is placed in columns, or shafts, which center all body-extensions in such positions that those centers are the location of Magnetic stillness, and, as such, are of zero electric potential.

Next in our consideration is the brain flesh which surrounds the thought-rings of the central "switch-board" of the entire bodily nerve system. This, with the bone-marrow which assists message-carrying intercommunications between the Creator's Mind, and the imaged-forms He is creating, constitute quite a new science in itself, which is more needed by the medical profession than by our layman students. For this reason we cannot devote as much space to this vital subject as we would like to do.

Our present purpose is served by exemplifying the fact that every illness of the body is made in the image of one's thinking and the actions which follow mental, or sensed-decisions. We can retain our body normalcy, or we can easily damage it, by unbalancing our thinking and our normal way of life. If, therefore, one is ill, he should first look to the cause of it in himself. If people only realized that a happy body cannot exist without a happy state of Mind there would be no "mental cases" or insanity, much less headaches and other ailments.

There are two other shafts of zero potential besides the arterial and spinal ones which center the chest. These are the shafts which convey food to the stomach and the one which carries air to the lungs. The membranous flesh, which constitutes these tubes, is not like a plastic container. It has great conductive supersensitivity. It performs two separate and seemingly impossible opposite offices, for it both insulates that which should be insulated and conducts that which should be conducted.

One is very often mistaken in assuming that a badly upset stomach is due to food unwisely eaten when the fact is that unhappy rhythms have been the cause. The reaction has not alone been confined to the digestive functions. Its main cause may lie in emotional disturbance, business worry, worry of conscience, fear or many other states of unbalance. Even the food one eats should be "happy." It should be cooked with love and eaten joyfully, and there should be a joyful realization of love in one's deep breathing and exaltation during the process of taking food into one's body. The food you eat becomes blood and flesh of your body, and the manner in which you eat it, and your mental attitude while eating it, decides your blood count, the balance between acidity and alkalinity of your digestive machinery, and your entire metabolism.

These four great body shafts have much more "responsibility" than just performing physical, mechanical purposes. Please realize that fact. Your typewriter is a machine which will supposedly operate just as perfectly if you are angry, but even your typewriter could register your anger and be damaged by it. Whatever your mental condition is, your body condition records and reflects it as truly as a mirror equally images an angry you or a happy you.

Your Mind is YOU and your body is the record of your thoughts and actions. Your body is what your Mind electrically extends to it for recording.

Future generations should learn to think in such terms. They should discard the present concept of Mind and matter, and substitute Mind, thought, and action in place of it. Some day man will fully comprehend that *matter is but the motion of thought.*

The substance of this chapter is for the purpose of making you realize that your Mind reaches to every part of your body, not only your arms, legs, fingers and toes, but to every microscopic cell in your body.

You can instantly order your arms or fingers to obey a command from you, because large enough nerves connect your "switchboard" with them. There is not one cell in your body you cannot reach, however, with orders from your Mind. You may have to concentrate long and often to do it but you can do it. A simple experiment to prove that is to look concentratively upon one part of your body, such as the end of one finger. Demand of it that warm blood shall collect there, and it will, after a very few efforts. Many have done this. Another experiment for convincing you will be to order your ears to move—and they will in due time, just as a Paderewski orders his fingers to do fifty times as much as you can with yours.

Now apply this to your headache. First making sure that you

are "in tune with the infinite" and not out of tune with it, and that you are not violating any law of Nature which continues its cause, your headache must disappear. It cannot do otherwise. Nature is normal. It balances all unbalanced effects if it is given a chance. However, it is useless to try to cure an unbalanced condition mentally unless the unbalanced *cause* is first replaced with the joyous rhythms which approach the God-Mind-ecstasy.

Now that you know where your Mind is in relation to your body, you have the basis of what Jesus knew when He extended His balance to the unbalanced thinking of other bodies, and gave them His balance. Unbalanced bodies cannot remain unbalanced if one's Mind is balanced and joyous to the point of ecstasy.

It is our earnest hope that each of you will now more fully comprehend the meaning of the command to seek the kingdom of heaven *within* you. It is also our earnest hope that the medical profession will apply this knowledge to every patient and not treat his physical unbalance alone, while allowing its mental cause to be ignored.

The physician is the logical Mind-healer for he has knowledge of *effects* which the Mind-healer cannot possibly have. One cannot go to God abstractly. You must know what you wish to ask for and know how to do your part in working with Him.

CHAPTER X

The Material Nucleus of the Atom

I

THE FOREGOING CHAPTERS have gradually prepared you to comprehend that the Rutherford-Bohr atom, which is symbolized in Fig. 58, has no place in Nature. This chapter is to tell why not one single fact of Nature supports it in structure, nor justifies it in principle. We have stated before that this electric universe has been constructed entirely by the electric wave, therefore, any theory of motion or matter, which is *outside* of electric wave mechanics and construction, is invalid.

If human reasoning by the greatest of the world's thinkers built this impossible concept, there must be something very basically lacking in knowledge of our universe to make it possible for thinking people to form such a concept. And there is something basic, many things in fact, *which sense-observation alone can never tell to man.* We will innumerate these also, step by step:

1. Nowhere in Nature does motion, in any three dimensional mass, revolve around a common center. All mass is made up of pairs of ring units, which are joined together in parallel planes to create hemispheres. Each ring has its gravity center and each mass has its dividing equator. Adjoining rings which form spherical masses cause gravity centers of these rings to seemingly become gravity shafts. All motion, in every hemispherical mass, spins in parallel planes which are 90 degrees from the gravity shaft of that mass. *This means that all motion in any spherical mass spins around its gravity shaft in parallel planes and not around a common center of gravity, which supposedly controls the whole mass, but actually controls motion in the plane of the*

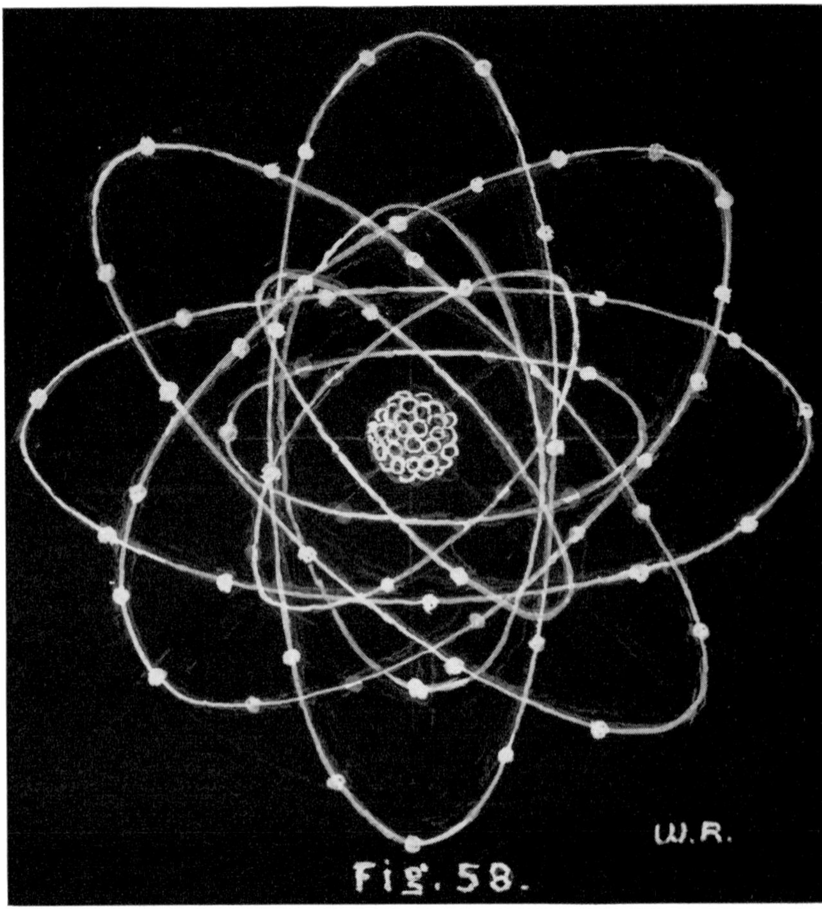

The Rutherford-Bohr atom.

equator only. See Fig. 59. This also means that Nature is bilateral and not radial.

All mass is divided by an equator, and the four pairs of rings which constitute that mass, are arranged in plane layers on either side of that equator. All pairs are sex conditioned by the divided spectrum, which place the red units of each pair in one hemisphere, and the blue units in the other hemisphere. The red units of a mass do not revolve across a common center of gravity and spin through the blue side to complete its orbit, as the units of the Rutherford-Bohr atom do, as shown in Fig. 58—nor do the blue units spin through the red. That would be as much against nature as it would be for humans to change their sex during every daily cycle.

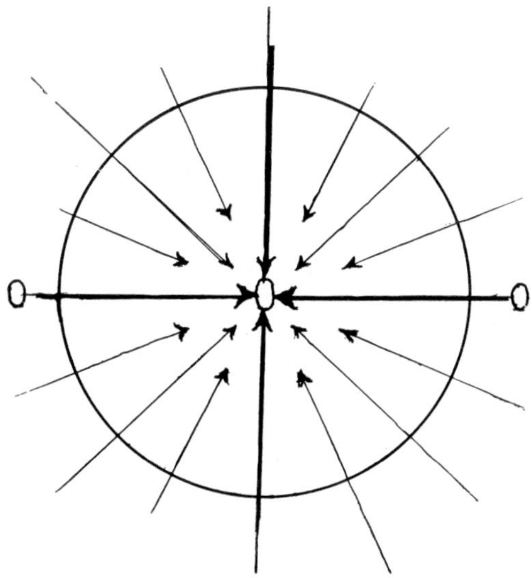

FIG. 59. Illustrating the radial universe concept.

The invisible universe is based upon the cube in relation to its control of motion. The stillness of gravity, therefore, when its points are extended to appear as a shaft, is always at an angle of ninety degrees from its electric thought-ring extensions. Never anywhere in Nature does any electric effect reach over to another center than its own. The three inner intersecting planes of the cube will not allow of such a crossing. If you examine the cleavages of a cube crystal you will not find them radiating from a center. The senses have been too much affected by the symmetry of light radiations from a star, or from the hexagonal radiating arms of a snow crystal, and many similar effects, which the senses too readily accept. It is time that we at least begin to doubt the evidences of our senses in making hasty obvious conclusions.

Another cause of confusion by the senses in this respect is the fact that all planets in a system, revolve around a common center of gravity. These are rings of motion, however, which spin around holes in which their controlling center is located. Every part of the mass of every planet, however, spins around its own center which is on the gravity shaft, and always in *parallel planes,* not intersecting ones, which would be necessary if all parts moved around a common center.

2. If you look at an electric current in a vacuum tube you will see parallel rings extending from the cathode to the anode, like buttons strung upon a thread. See Fig. 60. Each of these rings is controlled by its own center of gravity in its own plane. Each ring is a unit of a cycle and it is necessary for pairs of cycles to unite to form a mass, or for four pairs to unite to form a sphere. *When a sphere is thus formed it is as bilateral as all things else in Nature are bilateral.* It extends its right arm and its left one to form a shaft. Each point of gravity in that shaft is also the fulcrum of a lever, which its ring is. The fulcrum is still, and in it is all the power which its spinning lever-ring expresses, but *the energy thus expressed is not in the moving ring, it is in the fulcrum of gravity which centers it.*

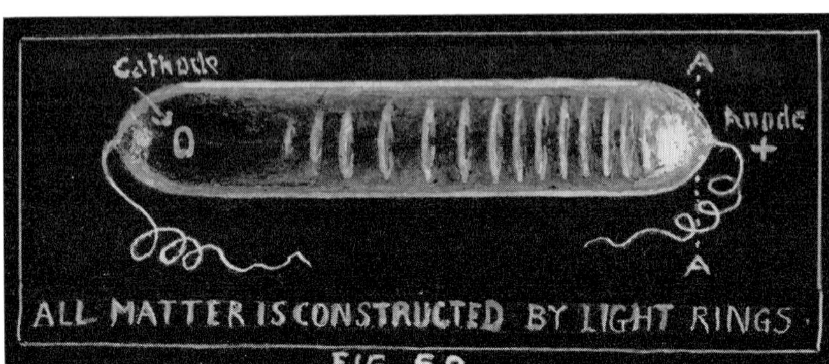

Electricity records Mind-thought in rings of light which spin around a Magnetic center of still, invisible light. All motion in all the universe spins around Mind-centers which become shafts as motion extends.

If you now look at the wave with this thought in your mind you will, likewise, see pairs of rings strung on the wave shaft of gravity which extends, as thought-rings of motion extend, to center and control the motion of each ring from that center, and to become a shaft of gravity for each mass which is formed by the amalgamation of ring pairs. See figures 9 to 29.

By a careful examination of the construction of an electric current, and an electric wave, there is no part of it in which orbits of units intersect the planes of the orbits of other units, or of masses which are formed by the union of pairs, nor do any of them share their gravity centers with other units. As there is no precedent in the electric current or wave of Nature to justify the multiplane concepts shown in Fig. 58, and as there is no precedent in the electric current or wave to justify the presence of material nuclei in the atomic units which constitute mass, such an impossible and unnatural concept should be expunged from human thinking.

3. Let us now apply this principle to the copper wire which carries a current. The wire is still. It does not move. Motion spins around its surface in rings, which are centered by holes where gravity sits and rules each ring. The wire is horizontal but motion is vertical in the plane of its spinning. The wire is a series of gravity centers which make of it a seeming shaft. The wire-shaft does not move yet it performs the tremendous work of lifting thousands of tons, and driving thousands of engines and other machines. And it could easily kill a man if motion, which encircled the wire, collided with the man.

Our senses very plainly, and truthfully, tell us that the wire is not moving, but if a speeding bullet traveled along the path occupied by the wire, our senses untruthfully tell us that the speed of the horizontally moving bullet killed the man, instead of the vertically spinning motion. *It may be a long time in man's unfolding before he can differentiate between what his Mind knows and his senses believe they know, but new thinking must someday come to man which is based upon Mind-knowing instead of body-sensing.*

4. Let us now bend the copper wire into a semi-circle to symbolize a wave, and send a current through it. We have now curved our gravity shaft and our planes of motion, which spin around it at angles of ninety degrees, and are no longer parallel.

They all point in the direction of the center of a circle and our senses tell us that that center is the common center of gravity for all of those converging radial lines. That is another misleading illusion which our senses believe in implicitly, but is not true to Nature. We must give space enough to very briefly tell what the Mind knows in respect to the curvature of gravity, which the senses cannot sense.

To help visualize this let us first disturb the horizontal plane of calm water and bend it into a semi-circular wave, as we bent the wire. The electric compression pressure registered fifteen pounds per square inch before the water was disturbed. It now registers less above the surface and more in the trough of the next wave. The gravity shaft curves as electric pressures curve. Our senses tell us that the wave shaft is moving up and down. Also they tell us that they are moving right and left. We speak of the speed in which light waves travel, which is proof that we believe that they do travel, when they but move up and down. Light waves *reproduce* at great speed, but light does not *travel* at any speed.

These are the things our senses tell us but our Mind knows that the only direction of motion is the electric direction of circles, which spin around gravity shafts to cause effects which our senses thoroughly believe in but which are not factual. *Light does not travel. It only appears to, and that appearance is limited to the boundaries of its own electric wave-field.* We do not yet know the universal radar principle which repeats every happening anywhere to everywhere. Our senses are mightily confused between zero *cause* and dual *effect.*

5. The greatest, and most visibly evident proof of the fact that the universe of suns and earths is made up of pairs of rings, and that they disintegrate by throwing off pairs of rings, is to look at the greatly magnified pictures of such planets as Jupiter and Saturn, and such ring nebulae as Lyra and the others on that same page. The great telescopes very clearly show the parallel wrinkles on Jupiter, which make its surface look like heavy corduroy.

The fact that all such planets and suns oblate proves that the rings expand in their own planes and not radially. If a planet expanded radially it would still be a sphere, no matter how large it grew, but all through the heavens you can see planets and suns expanding in planes which are parallel to their equators. This they continue to do until great holes are bored right through them where their gravity shafts exist to still control them until their motion ceases entirely. The Dumbbell Nebula is an excellent example of a sun thus expanded until a very large hole has bored right through it as it slowly explodes. In all probability part of it will regenerate a new sun in its center as the Lyra Ring Nebula, and countless others have done.

6. It now remains to explain why gravity does not express itself in radial lines from the surface to the center of a sun or planet, as diagrammed in Fig. 59. With that knowledge you will better comprehend why Nature will not tolerate such intersecting orbital planes of motion as the theory of the nucleal atom demands, as shown in Fig. 58.

There are only two spots upon the surface of a true spherical sun where a stone, dropped upon it, will continue its electric journey to its very center. Those two spots are the two "magnetic poles," which are the ends of the gravity shaft upon which the whole mass turns. Likewise, there is but one plane, where a stone, dropped anywhere upon it around the entire sun, would continue its electric journey to the sun's center. That plane is the equator which divides the two hemispheres.

Do not lose sight of the fact that suns are made up of pairs of rings which parallel their equators, and also do not lose sight of the fact that each ring is an effect which is centered by its own controlling cause. Likewise, bear in Mind that gravity is omnipresent and does not move.

With these warnings we are now ready to drop a stone on the earth a thousand miles from the equator. (We will ignore, for the purposes of this demonstration, that the earth is not now a true sphere as it once was, and, for that reason, the gravity poles do not coincide with the pole of rotation.) The stone now curves toward the pole as it is dropped, and its direction toward its centering gravity shaft still continues to curve away from the earth's gravity center, instead of directly toward it as our senses have long told us it would do. See Fig. 61. (This oblate earth actually has two gravity centers, but we will ignore that also.)

We drop another stone another thousand miles farther north, and still another with the same result. Each ring plane owes its allegiance to its own center of gravity in its own plane, but as there are countless billions of rings within rings between the surface and the gravity shaft, the line of gravity control will forever curve, until the last stone cast will be nearer the north pole than the earth's center, toward which it was originally headed, and where our senses told us it would go. We print Fig. 61 here to illustrate this fact of the curvature of electric potential pressures, which gives the impression that gravity itself is curving.

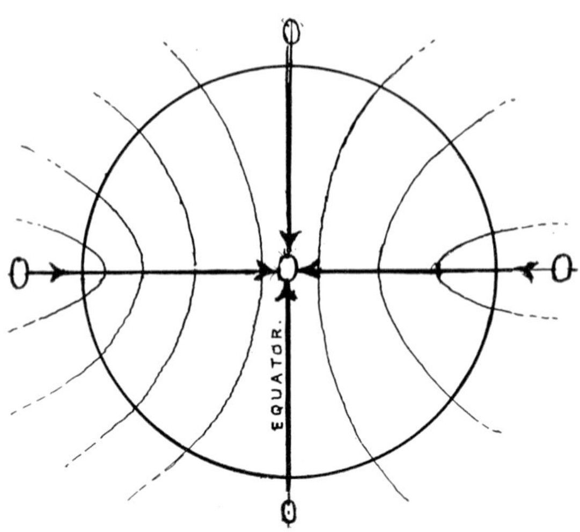

FIG. 61. Illustrating the bilateral nature of matter and the seeming curvature of gravity.

You are probably familiar with the bar magnet experiments with iron filings which curve from pole to pole. These curved lines are known as magnetic lines of force. You may also be familiar with the curved lines from the earth's equator to its poles, which the text books call the earth's magnetic field. These curved lines are in no way related to magnetism. They are but lines and planes of electric pressures.

Fig. 61 will graphically diagram this effect but figures 17 to 29 will help to clarify them still farther by closely studying them all, especially figures 17 and 18, and comparing one with another. When you have done this you will no longer accept the possibility of such an effect in Nature as the nucleal atom such as shown in Fig. 58. Also you will more clearly comprehend the marvels of the advance in electronics, which is making such marvelous strides against the resistance of sense-believing in things which the Mind knows to be otherwise.

II

Perhaps the greatest blow to the nuclear atom theory is the fact that it claims that each consecutive atom in all of the elements from hydrogen to uranium changes its substance by the addition of one more electron revolving around its nucleus, and a balancing number of protons in the nucleus. For this reason it numbered the elements from 1 to 92 at the time the theory was conceived. That meant that hydrogen had one electron revolving around its nucleus, helium had two, lithium had three etc., up to uranium which had 92.

No more inconsistent and fantastic concept could possibly have been formed than this, for it has not the slightest resemblance to Nature's processes, whatsoever. It assumes that hydrogen is the first element of Creation, instead of the 18th, and gives numbers to isotopes, which are split tones, in the same measure that it gives to full tones. It has not allowed for the possible discovery of new elements for which it could have no numbers to give it. This embarrassing situation actually arose when it was discovered that hydrogen was not the first element. The discovery of deuterium and tritium in the hydrogen octave threw that whole theory in the waste basket, but science met that danger by calling the new tones isotopes. Even then they had no numbers for these isotopes although they had numbers for the 49 other isotopes which you can see in Fig. 5 as they arrange themselves on either side of their wave amplitudes in the 6th, 7th, 8th and 9th octaves.

This unnatural concept destroys itself by its impossibility to conform to the chemical law of valence, which gives equality and multiplication and potential to mate-pairs in each octave, instead of 92 consecutive multiplications which ignored the octave-wave pulsation principle. Let us analyse this briefly to throw light upon this impossibility of Nature.

All through this book we have shown how God divides every effect of motion into equal pairs in nine consecutive waves. It is the most basic principle of Creation, for polarization is founded upon it. All through the nine octaves all full tones are divided into *four equal mate-pairs*. The first three of these pairs are equal and opposite atomic ring units, which increase their potential from 1 to 4. The fourth pair in each octave is united as hemispheres, such as carbon, and is the maximum potential of its octave-wave.

The law of valence recognizes the progression of multiplied potential and classifies these mate equalities by giving the first pair a valence of 1, such as lithium and fluorine. It gives a valence of 2 to the second pair, such as beryllium and oxygen. The third pair has an acknowledged increase of potential, which gives it a valence of three, as in boron and nitrogen, and the fourth pair—the carbon hemispheres, have a valence of 4. We will point out the havoc this concept would play in science if we tried to uphold the principle by endeavoring to fit it into Nature's scheme Lithium, for example, is given 3 electrons and its equal mate—fluorine is given 9. Nine electrons to 3 overpowers one of the mates and destroys its polarization equality. Much worse than that, however, its 9 electrons make fluorine, which is a very light gas with a below zero melting point, superior in potential to carbon, which has only 6 electrons, even though it has the great density which requires a temperature of 3,600 degrees to melt it.

Carrying that comparison farther we find that bromine is given 35 electrons and iodine, its equal mate, is given 53, which is nearly nine times the potential given to carbon. It is not difficult to compare the potential of iron to lead, yet the number of electrons given to lead is 82, or nearly three times the number given to iron. To carry this analysis still farther is useless, for the comparison multiplies the fault of the principle to the point of great embarrassment.

It was bad enough in the early days of research when Mendeleéf gave atomic weights ever-increasing potential in the same manner, for he gave all of the dying elements increasingly greater atomic weights than the living ones. As an example of this inconsistency the soft, dying element lead is given an atomic weight of 206, or over sixteen times the weight of the carbon atom, which is only 12. In this concept of atomic weights iron is given as 55, or four and a half times the weight of carbon, and tungsten is given as 184, or about fifteen times the weight of the one element, which is the most fully alive, most dense, and has the highest melting point of all of them.

It was as inconsistent to assume that all of the elements were progressive, or sequential multiples of hydrogen, as to assume that each successive element above hydrogen owed its characteristics to one more electron and proton added to those it theoretically possessed.

In neither of these two cases has the octave-wave been considered as starting from zero at cathode and colliding as maximum potentials at anodes. More conspicuously in error, than these two above mentioned, is the giving of electrons to the orbits of the inert gases, which are zero in every case and can have no atomic weight at all, for there is nothing to relate weight to in them, nor any complexity beyond the four undivided rings of their own construction.

III

The final step in proving that the nucleal atom is outside of Nature is the fact that such a group concept is, necessarily, three-dimensional, which means mass, and Nature does not begin three-dimensionally. It is first two-dimensional, for Nature is made up of single plane units, which are two-dimensional. *The moment two-dimensional units are divided into pairs, and unite to become mass, the three-dimensional bilateral universe appears.*

Let us very briefly review the process of Creation in its initial stages. *Mind-Idea must first be conceived.* The conception of Idea is registered in the inert gas ring, which is forever the seed-record of that Idea. That first undivided ring is two-dimensional. It has length and breadth, but not thickness. A ring is the only form in Nature that can be unitary, or two-dimensional, or balanced in itself, for its extensions are all in one plane. The moment that anything is added to a ring it must be added to on *both* sides to balance it. *It cannot be added to on one side for Nature does not create hemispheres, it creates spheres.* To add one ring on each side means the extension of a Mind-center to three Mind-centers, or a Mind-shaft to balance and control three instead of one. To add successive pairs means to extend the mass to spheres, and that means three dimensions, and it also means the creation of a bilateral mass. A three-dimensional universe is but the result of polarization. Polarization is not an extension of motion. It is an extension of omnipresent stillness into omnipresent stillness. *Motion is a lever which must have a fulcrum.* As motion extends there is always a fulcrum everpresent to center it. A gravity shaft, which results from polarized division of stillness into moving pairs, is not a direction of motion, nor is it a part of Creation. It is merely an identifiable Mind position in the zero invisible universe.

The Creator follows conception with action and reaction. The inert gases divide and extend. Motion is then created for the purpose of giving body-forms to Idea. Divided and extended motion is expressed in pairs of light rings. Divided light-rings must necessarily divide the white Light of Mind into the tensions of the color spectrum. That means the red and blue of a bilateral universe which cannot exchange its sides, for each is interchangeably sex-conditioned. The sex strains and tensions are set up in the immovable rings of the inert gas of each octave, which is divided into pairs. Motion, tension, strain and heat begin with the extension of motion, in ring pairs, from its conception in the four recording rings of each octave. These four rings of the inert gases are within each other in the same plane. They are, therefore, two-dimensional. *The moment they divide and extend into unit pairs, to simulate body-forms, mass begins and the three dimensional, cube-sphere, bilateral universe ap-*

pears. In other words neither God, nor man, create three-dimensional body-forms before conceiving the Mind-image of that form. Mind-images are thoughts. Body-images are actions, and actions are always in pairs. Thought-concepts are two-dimensional. Actions are polarized extensions, so actions are, necessarily, three-dimensional.

The Bohr atom is a three-dimensional result of action which precedes its concept. Such a concept is entirely outside of Nature for it has no precedent in it, nor is it a part of the electric process which constructs mass. Mass is an extension of many thought-concepts into many planes, and a simulation of extension of the one Mind-center into a gravity shaft.

SUMMARY

God—the knower—is non-dimensional.
God's thinking is two-dimensional.
God's creative actions are three-dimensional.

The nucleal atom theory begins with three-dimensional mass. It conceives groups of neutrons and protons revolving about and around each other on both sides of the spectrum. Each neutron and proton is conceived to be a three-dimensional, spherical mass in itself. In combination they constitute a larger three-dimensional, spherical mass. Three-dimensional beginnings are impossible. The electrical Nature of this universe demands a division from one plane located in a cathode. *Polarization is the result of that one-plane division. Balanced equal and opposite pairs thus come into existence. Every united pair becomes an anode, which means a hot bodied three-dimensional mass, which is straining explosively to return to its non-dimensional beginning, which it does, but in doing it it leaves the record of its motion in two-dimensional thought-forms in its inert gas.*

PART III

PRELUDE

God-Consciousness and cosmic awareness of the Light of the divine Presence within every man is the next step in the spiritual nature of man. Thousands—and then more thousands—are beginning to comprehend that Inner Voice of one's own Mind coming from within each man. The reason for that is because the human race is still in its intellectual infancy. The Dawn of Mind-Consciousness is too recent for all men to THINK with their Mind.

Man-in-the-mass still senses with his body. His desires are still dominated by his senses. His concepts are still sense-based. He hurts himself with his own acts and calls it evil. He conceives a personal God with vengeful human emotions.

All that Jesus told the world about the unity of God and man, and the location of the Mind-kingdom within every man, has no meaning except for the very few illumined ones and the very many who now so deeply desire that illumining.

Yes—God-Consciousness is slowly infiltrating the human race as man knows more and more how to THINK with his Mind instead of SENSING with his body. We, who know God in us, are desirous of illumining the path for you to find that Light of Inner Knowing—and when you do find it, you will then be able to say—with deep knowing—I and my Father are ONE—and you will know all things.

CHAPTER XI

The Bridge Between Man and God

I

"The last enemy that shall be destroyed is death."
1st. Corinthians, 15–26

THE FIRST and last great mystery of Creation is its Creator. Second to that mystery is the eternally asked question: "From where did we come to live and where do we go when we die?" This very simple question is quite easily answerable to the spiritually unfolded one who knows that which he cannot *see*. It has always been unanswerable to sense-dominated man to whom the invisible universe where God dwells and rules, has no meaning, for he has no knowing beyond his sensing.

There is a very definite bridge between God and man. It is invisible in the vacuum of the zero stillness of God's kingdom, but it has a visible link at every point where that bridge touches the shores of the action-universe of motion. That is something which every man can understand, for he can hold the symbol of that great REALITY in his hand in the form of a simple seed. *He has never known, however, that in the imperishable seed within that perishable acorn, which he holds in his hand, is the whole answer to where the oak tree comes from to live, and where it goes when it dies.* That is something which the physicist can more readily understand when you tell him that the imperishable, invisible seed within that acorn is an inert gas, or a combination of several octaves of inert gases.

We again refer you back to Fig. 5. You will note that an inert gas, marked zero, begins and ends each octave. The nine inert gases are the shores of the visible universe, where the invisible bridges, which link Mind and motion, touch the moving-action universe. There are nine of them because the bridge has nine parts, which we might call entrances to mortality and exits to immortality.

Man has always thought of his periodic entrance into mortality as an entrance of his Person—his being—his individuality, into life. And he has always thought of his body exit into his eternally living Self, as death. That is the thought which has always been man's great enemy. That thought of death shall be destroyed by knowing that there is no death in Nature to destroy. There is naught but eternal life in all this universe. *There is naught but Mind at rest, and Mind-thinking in motion, in all Creation.* Such a concept must gradually change as unfolding intelligence in man gives him higher knowledge of REALITY rather than its loyally believed in *simulation,* which motion produces.

Eternal Identity is never created. The IDEA, which man is, is never created. It always retains its dimensionless zero in the cathode of itself. That body, which man thinks of as man, is a moving light-image which is projected from the dimensionless point of man's eternal Self. The projected light is mirage. It is a series of repetitions and reflections projected through invisible cube wave-field mirrors. It has no more reality than cinema bodies have. It has no more knowledge or intelligence than they have. These bodies cannot even move of themselves just as cinema bodies cannot move of themselves. Some day every man must realize that he is MIND ONLY. What he now thinks of as himself is but Mind-thoughts electrically projected in two seemingly opposite directions in order to provide pressure mirrors of invisible light to act as screens for his light projections. He must learn that his body is but a formed, moving image of his Mind-imaginings. Also, *he must learn that there is no death in God's universe, there is naught but perpetual rebirth into an endless repetition of rebirths.* Every opposite of every pair is born, and becomes its opposite sequentially. God's words, as given in His Message of The Divine Iliad, expresses this thought in these words:

"Know thou then that I alone live. I do not die, but out of Me comes both seeming life and death.

"Life is but the inward flow of My thinking's divided pulsing, and death is its outward flow.

"Know thou also that the divisions of My thinking are but equal halves of One; for I again say that I am One; and that all things which come from Me are One, divided to appear as two."

It would be well to simplify our thinking as to what Creation really is by putting it into simple words. *Creation consists of the invisible universe of Mind, which Jesus called heaven and man calls space, and the visible universe of motion, which man calls matter and substance.* There is a vast difference, however, in the meaning of the omnipresent heaven, which Jesus referred to as being "within" man and within all things, and the meaning of heaven which man holds as being a place up above and apart from the earth. For ages man has thought of heaven as a place above the earth where dead bodies and souls ascend. Literature and art are replete with symbolism which very vividly pictures that primitive concept. If we but enlarge our concept of the meaning of the word heaven to be that zero omnipresent vacuum which is everywhere within and without all things, as symbolized in Fig. 6, then the ages-long idea that we go to heaven when we die is literally true. The difficult thing for man to imagine, however, is the fact that he returns to "heaven" non-dimensionally, just as the dimensioned oak withdraws its huge dimensions into ultra-microscopic nothingness. That is man's difficulty. *If he would but realize that his body is but the projected image of an Idea, and that Idea has no measure, it would help him.*

We must bear in Mind that the primitive pagan idea was that the earth was all of Creation. Man then thought of heaven as surrounding the earth. Heaven even had a limitation with holes punched in its boundaries for stars. Pagan man had no conception of limitless space and billions of suns and earths. He thought locally. Even his God was for this earth alone, and even for certain races alone. The Greek gods were man-formed and had human emotions and desires. That pagan, man-formed God still persists, a God made in the image of man, a God with arms and legs for which He has no use in a limitless zero expanse, a God who could be angry and wrathful with humans, who could not possibly be omnipresent because of his objective limits. The time must come when man stops praying to a God "up there in heaven" *outside* of himself, and talk to God *within* himself. *The day of the brotherhood of man, with peace on earth and ecstasy in the heart of man, will never come until man awakens to an awareness of his own divinity, and finds God in the Light of Mind, which he, himself, is.* May that day be soon for the many who are ready, and may they hasten the day for the still pagan-minded by awakening the Light within them.

We must become more familiar with the office, purpose and construction of the inert gases in order to have a more definite knowledge as to our own immortal identity in that omnipresent heaven. To acquire that knowledge and comprehension the very first step toward it is full comprehension of the invisible bridge between heaven and earth, and the visible entrances and exits to it through the nine inert gases. Full comprehension of the interrelation of figures 6 and 7 is a necessary prerequisite to comprehension of further stages of God's ways and processes in this respect. Next in import is to become thoroughly familiar with the nine stringed instrument upon which the symphony of Creation is perpetually being played. Fig. 5 has been prepared for just that purpose. At the beginning and end of each string is an inert gas.

Each inert gas is constructed by four rings in one plane, centered by a hole which is the invisible Mind-Source of those four rings. At the very center of that hole is a point of stillness, within which is lodged all of the life, energy, knowledge, idea, and the other qualities which are a part of the God-Nature in non-dimensional qualities. Every point in all the universe is like that point, but we are concerned with but one of them now, for that is the point where the oak tree draws its power to express the oak tree idea in form—or which any other unit of Creation has chosen to draw its identity and power.

That omnipresent point from which you have issued your body, is the same point at which the visible, mortal you becomes invisible and again assumes immortality. That one point in all this universe, controls your every movement from your first one many millions of years ago, to the last one which consummates the idea of man as expressed by you. That one point is *your* Soul of the universal Soul. It is *your* Mind of the Universal Mind, as one unit of Creation. You are ONE with that Mind—in fact you *are* that Mind. *Your body is the thinking and imagining of that Mind.* The eternal YOU is that center, and your eternal BODY is that series of four rings of fluorescent light which is eternally wrapped around that centering Soul of the eternal YOU. Those four rings are eternal records of you. They are as immortal as you are immortal. They are the micro-film of you *yourself.*

Do you begin to grasp the stupendous significance of that idea? If you do grasp it you have become aware that neither YOU nor YOUR BODY can die, for that eternal micro-film of you will forever be projected into three dimensional enlargement after every rest period between action and action. *If you really do comprehend it you now know that your Mind cannot die, for your Mind is ONE with God's Mind, and that cannot die.* Likewise, your body cannot die for it is Mind-thinking, and the *record* of Mind-thinking is enfolded forever within its Soul-seed and recorded upon those four rings which surround it exactly as a Beethoven symphony is recorded upon the rings of a phonograph disc.

YOU are eternal, but your body is eternally repetitive. Your Mind never sleeps, for it is changeless, but your Mind-thinking rests from Mind-thinking in cyclic intervals between its pulsations. Your body thus sleeps every night and becomes unaware of body-existence. It must do so for it needs replacement and repair. You are quite familiar with that effect. It is quite understandable to you because of its oft repeated occurrence.

For the very simple reason of your newness of unfolding intelligence as man, you have not yet grasped the idea that the end of a body cycle, where the body has completely worn out, or where its growth has been short-circuited by a disease, or bullet, or from falling over a cliff, is just another interval of sleep and body-replacement by light-projection from its seed-records. *That fact you do not yet understand, but you will if you meditate upon it sufficiently to let the knowledge we are giving you sink deeply into your centering Consciousness instead of holding it superficially in your senses.* When you do fully understand this fact you will then know that what you have made yourself to be is what you, yourself, have electrically recorded upon those four rings which surround your Soul-Identity. Whatever is recorded on those rings is perpetually being re-enacted by you every second of the day. If you look into the mirror you will see what kind of person you have made yourself become today because of your thinking and your acting a million years ago. You are the kind of person you have desired to be. *You are the sum-total of your own desires.*

Now look at some person who is an infinitely greater Soul than you. What is the difference between each of you? *It is a difference in desire only.* If a greater one than you inspires you to be like him, and you desire to be like him, you can be by recording that desire upon your Soul-record as a mental Image, and that Image of your Mind-desire will be projected back to you and make you like unto it.

Let us again look into this new deep well of Mind-thinking and consider that you are ill of body because you have made it so by desiring to make it so. If you will but realize that you are recording the pattern of defective thinking upon your Soul-record as well as right thinking, and desire instead that the normal recordings of long ages of building a normal body will regive you a normal body NOW if you will but let it. Your illness will be voided and replaced with the normalcy of the light-image of your ages-long building of your identity. If you will but look again at Fig. 6 and realize that the omnipresent, omniscient, omnipotent Mind of the invisible God Light, therein symbolized, has absolute rule over the projected light of His thinking, as symbolized in Fig. 7, you will more readily comprehend how it is that the invisible omnipotent, omniscient YOU can, and do, project balanced, or unbalanced conditions to your body from your Self to *create yourself in your own Image.*

All things created by God are created in His own Image, therefore, God's creations are balanced. As you slowly arrive to the point where you know and feel God's Presence within you, your creations will have in them the balance and masterliness of God's Creations in the measure of that awareness.

What is true of you is true of a whole civilization. Man-in-the-mass is not yet aware of the Presence of God within him, and that is why this civilization is being built in man's image—not God's. This idea is clearly expressed in The Message of The Divine Iliad in the following words:

> *"For I say that man who senses but clay of each in him is bound to earth as clayed image of earth.*
>
> *"Clayed images of My imagining who know not Me in them are but dwellers of earth's dark. To sensed man the doors of My Kingdom are self-barred by darkness until the Light of Me in him is known by him as Me.*
>
> *"Until then he is but moving clay, manifesting not Me in him while sensing naught but moving clay of him; knowing not the glory of My Light in him.*
>
> *"Wherefore, I say to thee, exalt thou thyself beyond thy sensing, Know Me as fulcrum of thy thinking. Be Me as deep well of thy knowing."*

II

To enable you to more fully grasp the profundity of this idea we shall give you simple and familiar examples which relate to it. Perhaps the simplest of these is to recall something you must have often wondered about. Let us return to the example of the seed which you hold in your hand—and meditate upon it. You have realized that the ultra-microscopic seed enfolded within its surrounding placenta contains within it the image of every branch, leaf and fibre of what will be a fifty ton tree a hundred feet high. You have recalled to yourself that no two leaves—or two millimetres of that tree—are alike. You have undoubtedly recalled a hundred wonders of that tree and asked yourself how it is possible for all of that great tree to be enfolded within a microscopic point. *What you have never realized in this respect, is that you are endeavoring to translate the unfamiliar zero universe of Mind-Idea, which has no dimension, into the universe of dimension which you are familiar with.* That is a very difficult thing for you to do at this present stage of Mind-unfolding in the human race. A close approach to it in human technique with light and photography, might help you to grasp it. We will try a familiar one.

If, for example, you see a twenty foot high picture in the cinema, you know that it is a light-projection from a miniature image. You can see the big picture with ease, but if you look for a face upon the miniature, which you can see with great detail upon the screen, you will not be able to more than locate its position upon the film-image. Let us carry this one step farther. Human ingenuity has devised a technique whereby micro-films can be made. Now if you look at it you will not be able to locate the position of a whole man, much less his face, yet it can be multiplied into distinct visibility from that invisible point, just as the invisible oak can, likewise, be multiplied into distinct visibility. So, also, can ear-splitting deafening sound become silent and locked up as silence in a box, awaiting your Mind-command to again project it into loud sound.

Let us take another step. Light is limitless but human technique and material is not. Let us assume, however, that it is possible to project a picture one hundred feet high from that micro-film. If you carry this thought into deep meditation you will understand how dimension itself folds up to zero in the seed-record of your Self as it almost folds up to zero on the micro-film. If we now return to the one hundred foot high, fifty ton oak you can more readily comprehend that dimensioned, complex patterned-bodies can unfold in great variety of detailed effect from a Mind-concept of non-dimension, but, likewise, they can refold into non-dimension.

By taking another step we can, perhaps, help you to comprehend this idea more clearly. In your imagination look carefully upon the large cinema picture and realize that there is one prime essential to its understanding that is not photographed there. That one essential is the IDEA of the picture. The IDEA is not photographed. *It is not there at all.* You cannot *see* it, no matter how carefully you look—yet you KNOW it as well as though you *could* see it. Idea is never created. Its *body* is alone created. Idea has no dimension. *It is only the projected image of Idea, which has dimension.* Idea is never *created.* It is but *simulated.*

If you have followed these steps with even partial understanding you will comprehend the description we have given you of the bridge between man and God which the inert gases provide, and the further description we will now add to it.

To help you with this further understanding turn back to Fig. 41 which diagrams one inert gas unit many millions of times magnified. The central portion of that diagram represents the Mind within motion. It is absolutely still at its center, which is the kingdom of heaven within every creating thing, which Jesus tried to tell man about. That is God's universe, but it is also man's when he becomes aware of it. *Until man does thus become aware of it he is but a mortal body.* Unknown to him, however, for the long ages of his unawareness, something within that invisibility, which centers the rings, is eternally reaching out to him (and to all creating things), which he at first feels as instinct. That is man's first bridge to God. See Fig. 48 and read the words printed within the zero area.

Long, long ages pass before that bridge, which reaches out to him from its dark, becomes illumined with its own Magnetic Light of Mind-knowing, which then comes to him as inspiration. Even then man may still not know that God is actually talking to him in God's language of Light. That spark of awakened Light in him may burn in him as genius-man for many lives before he finally becomes fully aware of his Oneness with that Light. *Then there comes to genius-man that great blinding flash of Magnetic Light, which engulfs him and takes him across that bridge from his world of motion unto its very stillness.* Then he really does hear God's Voice saying: "See me—Be Me—Be thou fulcrum of thine own power." Genius-man has then fully crossed the bridge from motion into that stillness of the One which opens the door to the eternal omnipotent and omniscient Light of Mind. Then, and not until then, can man say: "I and my Father are ONE"—*and know its meaning*. That is what Christ Consciousness means. It is the final goal of all men. This man cannot yet comprehend, for the human race is still in its spiritual infancy and the journey of life is long. (The five stages in the unfolding of man's long journey from primate man to Christ Consciousness has, perhaps, never been more fully explained than in "GOD WILL WORK WITH YOU BUT NOT FOR YOU.")

III

Human pens can but write words. They cannot write their knowing into those words for others who have not that knowing. A symphony can be as well heard by moron and animal as by the genius, but the Soul of the composer is not heard by one and is movingly heard by the other. To some of you these words, which we have just written, are but words—and will be for thousands of years hence—but not to all of you. There are some who will understand and be greatly moved. To all of you, however, we may begin your comprehension by opening the doors to the invisible unknown just a little bit farther by more fully explaining the office and purpose of the inert gases.

It will help you much in this respect if you will but use your imagination in relation to an effect in Nature with which very few are familiar, the effect of ultra microscopic smallness. Man-in-the-mass thinks in terms of bigness of dimension. His senses are geared to dimension. He sees thousands of objects for miles about him and for millions of miles out into space. His senses are geared to motion, also, and the many lights and sounds which are one with motion. His senses are, likewise, geared to extension. They see only the unfolding extensions which are projected from the seed. He does not even realize that what he views in big scale out there is enfolded within him so near to zero scale in motion and pattern that all of it occupies only a microcosmic-scale image in one's eye. Is this clear to you? Do you realize that the *bigness out there* is identical with the smallness within your eye? *If so, can you not begin to accustom your thinking to the fact that the entire cone of the visible universe is enfolded within, and extended from its apex by the mere spiritual mechanics of creative imagining?*

As your power of thinking grows to this magnitude you will then soon be ready to ask of yourself the following question: *If objectivity and dimension are but imaginings, what then shall we say about the motion which created them?* If you fully comprehend these facts you can, also, fully comprehend that the bigness out there is an extended radial enlargement of the zero smallness within you. If you will but add to this thought the fact that the small image is a recording of the other, you are coming closer to a comprehension of that great secret of Nature. Now we ask that you add one more thought to this picture. Add to it the fact that the large and small image are not two. They are one. The same light which created one created the other, and the rays of that projected light connect the two as substantially as light-waves connect the sounds of a man's voice to the tape which is recording it. In other words, whatever is seemingly happening anywhere "out there" in extended space is being simultaneously recorded within the zero universe upon the four rings of the inert gas series, which bridge God's knowing to His thinking.

Now add to this thought the fact that every thought and action taking place in the extended universe of motion is simultaneously taking place everywhere in it. If that is so—and your radio and television have long told you it is so—can you ever again think of the universe as any more than ONE? If you have now arrived at this stage of comprehension in your knowing you can complete your knowledge of the Cosmos by including in it the One Source of that extended image you now know God, and man, and all things else, as ONE.

That which is transient in motion is eternal in the translucent Light of the inert gases which divides the Magnetic Light of Mind to produce pairs of electric bodies. These inert gases record all effects of creative thinking upon their four rings as precisely as the rings of a phonograph disc records extended effects. If you fully understand this you can now comprehend that as the oak unfolds from its zero seed to extend into space, the oak tree patterns of old and new growth simultaneously re-fold into its seed. Every leaf and branch are as perfect in the invisible oak as they are in the visible one. The marvelous thought about this is the fact that the extended one can perish, but the seed-record is imperishable. Nothing can hurt or affect that seed even though its planet may perish along with its extended "mortal" image. In other words, God's thoughts are not transient. They are immortal image-forms of immortal Idea. They repeat themselves in Nature eternally, for that is what Nature is. Nature is a light-extension of the Light of Mind-Idea.

The next step for you to take in greater preparation for your Cosmic knowing is to translate the effect of the oak tree body to the body of man. The projected image of a baby begins by the unfolding of the eternal record of itself, which eternally exists in God's Mind-universe. All of the thousands of detailed forms of cells and body parts are but repeated extensions of eternally existent thought-conceptions. *If you will but realize their oneness in the light which they are, you will fully understand that the body of man is but a projected thought-image in motion of an eternal thought-image in God's Mind.* That being so, can you not now fully comprehend that man cannot die? The light which images him may be shut off from its projected Source for a time, but it will again be projected from its eternal Source until its purpose for seeming existence has been fulfilled. *That is what is meant by the verse from 1st. Corinthians which headed this part of our narrative.*

IV

At this point it will be helpful if we enlarge upon the idea above stated in regard to temporarily shutting the light of Mind from the projected image of Mind-imagining. We mean by this exactly what you mean by saying: "John Smith was killed by a bullet."

When the light is shut off in the cinema the projected picture ceases, but the REALITY of that picture still exists as IDEA, and can be repeated. The entirety of this objective universe of moving bodies is precisely alike in principle to the cinema. It is a series of electrically projected pulsings of changing thought-patterns with black rest gaps between every change of thought-pattern. The pulsings are so fast that ever changing patterns create the illusion of motion and all of its kindred effects. The projected pulsings are transient effects, which ever change. *The Source of those pulsings is eternal, and never changes.*

There is no motion in the Cosmic cinema. All of it is a radial projection from a zero fulcrum of stillness. It is the apex of an unborn cone. A radial projection from zero creates the illusion of a three-dimensional universe. *Its sequences of pulsings create the illusion of time. Your body is a Mind-projection of your Self.* We are thoughts of Mind being made in the image of our imaginings. The multiplicity of mathematics, as well as the density of concentrative thinking, are expressed in infinite variety and complexity "out there" on the other ends of light rays, but they are zero at their Source within them. That brings us back to John Smith whom you say, was *killed* by a bullet. It is true that a whole system of the growth of thought-patterns, which constituted the projected image of John Smith, was interrupted, but it is not true to say that he was killed, for John Smith was not "out there" where he could be killed. John Smith is a Mind-Identity. He is not one thought of The Cosmic Thinker, but is also ONE with THE ONE THINKER. Never forget that. John Smith and the Creator of this universe are ONE. That ONE cannot be killed.

Let us parallel this thought with other words. Instead of John Smith being killed by a bullet, let us say he was electrocuted. That means that his light was shut off by sending a current through his projecting wires, which was too great for them to carry. If we now substitute a body of tungsten for the body of John Smith, and electrocute the tungsten, it also would be short circuited, or "killed," if you prefer to use that word. As we described in previous pages, a tungsten wire thus electrocuted in a vacuum tube leaves a residue of the inert gas helium. That helium is the Soul-seed of the tungsten body. It is the Soul-record of that body which is its identity in a unitary universe. It is the zero of tungsten's multiplicity. That cannot be killed, nor can it be prevented from *reincarnating* its image within it. *This is a universe of eternal repetition of pulsing thoughts which are ever being repeated.* In the Cosmic sense reincarnation and repetition mean the same thing, for Nature is eternal, it is *endless and beginningless.* The idea of death—or of identity—or of being killed—is a conclusion of the senses of bodies. Such a thought is impossible for the knowing Mind of the Cosmic mystic.

V

It is necessary to know why tungsten becomes helium. There are nine inert Soul-recording gases, why helium? The inert gas for the tungsten octave is xenon. Why should tungsten not refold into xenon? The answer is that it does refold into xenon, but all of the inert gases are within each other and helium is the balancing inert gas of the nine. Xenon expands into krypton. Krypton expands into argon, then into neon until it finds balance in carbon. This opens new worlds for further explanation for which we have not pages enough here. We can exemplify, however, by having you open the damper pedal of your piano and strike any one note hard. If you do this you will also hear every other octave note, above and below it, sounding in their own frequencies. One of these notes will dominate over all of the others, however, and that is the mid-tone of all the octaves. That mid-tone is the crest of the universal wave. All effects of Nature reach a wave-crest. They must, or perish, as the 9th octave perishes through inability to create a wave-crest by the uniting of balanced pairs.

Carbon is the mid-tone of all the elemental octaves. The radar principle of Nature is also another world for which we do not have pages for, but carbon is the radial cube of the zero universe and the radar principle in projected effect is cube-dominated. To fully comprehend radar one must fully comprehend the meaning and purpose of Light of Mind and pairs of lights of body, and how the latter act as mirrors and lenses in the light-pressures which constitute this light-wave universe. After all, a Cosmic cinema must have its projecting machine just as man's projector must have its mirrors and lenses. As a fact of Nature, the cube of nine light planes of zero curvature is the Cosmic projector, and the equipotential curved light-pressures, which are the product of that projection, are its lenses which divide and multiply projected effect. Out of this simple Cosmic process the curved universe of motion appears from within its cube planes of zero curvature and the 0–1–2–3–4–0–4–3–2–1–0 light-wave universe of seeming motion appears to deceive the senses of the human race of such an early stage of unfolding Intelligence as ours, into believing that the projected mirage-image is the real body, which thinks and knows, and lives and dies. Early man never gave a thought to the fact that his body is but an actor in the Cosmic drama of Cause and Effect created by Mind-desire to simulate the Love-Idea of balanced equilibrium by simulating that idea in moving imaged-forms by light-wave projections.

It is regrettable that we cannot help to further clarify this thought by giving the several chapters which would be necessary for explaining the light-wave which constitutes this entire universe, and radar, which repeats its seeming divisions. This was intended to be a small book, limited to one purpose. It could very easily become a massive one, by exceeding its purpose. We feel, however, that one paragraph can be spared to explain why it is that science is confused about light expressing itself as both *wave* and *corpuscle.*

The basis of that explanation rests upon the fact that the wave belongs in its entirety to the zero of gravity, while corpuscles of matter belong to electric motion. Every octave wave is a gravity shaft around which corpuscular matter spins.

IDEA IS ONE, but must be made two to produce motion by interchange between the two. If you will bear in Mind that such divisions merely simulate Idea, and that Idea is never created, you will more fully comprehend why John Smith could not be killed. John Smith the immortal was never created. John Smith, the eternal Idea, is divine. John Smith, the moving image, is but a *simulation*. That brings us back to two statements made heretofore which can be more clarified right here by adding the above stated thought to them. We stated farther back, that a certain point in the zero universe became your controlling center, or the Source of your Identity. The zero universe is not divisible, however. Mind-thinking at any one cathode point in it is universally repeated in every other point in it. In other words, *your* identity in the Mind-universe is God's Identity. Your identity is universally enfolded within all other identities. You are one created unit thought-body of the universal Idea of man. *You are not one, you are all men—billions of them. You exist in them and they in you. Whatever you think you do individually you are doing universally. Your thoughts are every man's thoughts and theirs are yours. Every happening anywhere by any man or other body in Nature, simultaneously happens everywhere throughout the universe, but each happening in any body informs all other bodies of its new condition of effect at a light speed of 186,400 miles per second.*

We have also said that there is no death in Nature. These two statements lead to much confusion to those who are not yet permitted to know that which they cannot see. One of the thoughts which thus confuses is the thought which is expressed by many who are grieved at the loss of loved ones. These continually ask why—if there is no death—they cannot communicate with their loved ones. This would be quite self-explanatory if one would but think things through by assembling the above stated facts as a foundation to reason from. Let us do this together.

We will draw a simple, and familiar picture. You are sitting by the bedside of your husband. He is sound asleep. He has not the slightest awareness of bodily existence. You do not expect him to exchange thoughts with you, or to accompany you to town, or to inform you of anything. Your companionship with him has as surely ceased as though he was in his casket, so far as his body is concerned. It would never occur to you to employ a medium to see if she could communicate with your temporarily dead-to-the-world husband's body, as many do whose loved ones bodies are in a similar, but permanent condition. Those who do this do so because they feel that their companionship with their loved ones has ended. As you sit there, however, are you not aware of a companionship of the eternal Thomas who cannot die? Everything of a Mind-nature which has come from him to you you still have. They are a part of your identity. The memories of them which have helped you to become you are still companion to you. You see and hear his thoughts in his works around you, the house he built with you, the children he borned with you. What is the difference between that sleeping body and a dead one? *The only difference is that the dead body of Thomas cannot awaken, but Thomas, the immortal, is still living and becoming a part of you, and of all men every moment of all future ages.*

If you do not grasp this ask yourself if your own life is not being enriched by Beethoven, Wagner, Shakespeare, Walt Whitman and the Emerson you so deeply love the companionship of. Would you be the same woman without them? If so are they not a part of your identity? Again, if so can you say they are *dead* when you can have their companionship whenever you wish? Who can you say is dead, whose body once walked the earth to be an actor in the play of Love, which The Master Playwright is throwing upon the screen of imaginings to simulate His One Idea, and to help lay a stone in the structure of civilization as one coral marine animal leaves its calcareous skeleton with other calcareous skeletons to build an island in the sea?

Can you say that Laotzu, or Shakespeare, Socrates, Homer, Newton and Luther are dead when they are continuing to make scholars and cultural citizens each moment of each day? Can you rightly say that Jesus died on the cross and will come again, when He is already here transforming your life and ours, and countless others by His teachings, which you can hear and read as effectually as though He spoke them to you from the voice of His body?

Who are you, therefore? Are you not all that has ever been? And will there ever come a moment in all future ages in which you are not eternally living among eternally living mankind to make it a part of you? If you thus reason can you not see the great confusion which exists in man's minds, or *senses* rather, because of mistaking immortality for mortality? Can you not see, and comprehend, that material bodies and material values go hand in hand with early man, but as his spiritual nature unfolds from body-sensing to Mind-knowing of God-awareness in him, man gradually exchanges material values for Intellectual ones? Can you not plainly see that man is not in an intellectually advanced age as yet, despite the seeming wonders of his material progress, but, on the contrary, he is still in the Barbarian Age of man-killing? *He has not yet discovered his unity with man and God, for when man really knows his unity with all men he cannot possibly kill.* Our present civilization has been built on piracy. That which one nation has taken from other nations for their enrichment has impoverished them instead. *No man will kill another when he knows that he is killing himself.*

Is it not clear to you that man has not even known the meaning of love as yet, for love means the action-reaction of equal giving and regiving? These work simultaneously under the law, but when man *takes,* instead of *giving,* the simultaneity of *taking* in action-reaction acts in reversal of the law of love. Can you not, also, plainly see that the Mind-universe is still but *words* to our civilization as a whole, when only a small percentage is on the edge of its meaning? The world intelligentsia very commonly refers to the electrical awareness of the senses as "the conscious Mind," and to the God-Mind as "the subconscious Mind." Many of our great intellectuals still believe the *brain* to be the *Mind,* instead of being but a receiver, recorder and transmitter of states of motion. These also believe that each individual has his own individual Mind, which dies when he dies. Many of these actually believe that Mind evolved from matter.

How wonderful it is, therefore, to see that science is at long last, beginning to not only discover the existence of a zero universe, but has become aware of its mathematical necessity. The recent discoveries of the anti-proton and other anti-matter particles, is the first glimpse that science has ever had within God's zero universe. For many years the necessity for a zero in a quantitative universe has been considered as necessary as a fulcrum of non-motion has been necessary for the expression of motion. The apex of the universal cone is as far as one can go in that direction, but whatever of infinite extension there is in this universe is in the direction of its base.

The anti-matter discovery of the last two years will become a great branch of science as soon as the offices and purposes of the inert gases are really known, for they are the key which unlocks the long closed doors to God's zero universe of Mind-Light. It is not proper, however, to use the term anti-proton, in reference to the Soul-seed of any condition of motion. It would be just as inappropriate as the use of the term anti-man in referring to his Soul-seed. What science is really discovering is the reality of the non-dimensional cathode from which the electric divisions are extended into dimensioned and conditioned anodes. *It is, at long last, realizing that the action universe of motion must have a fulcrum which does not move. This will lead to the placing of energy in the fulcrum Source of this universal mirage of motion instead of in the mirage-extension where Science now places it.* When that day comes, science will first question the universal vacuum for CAUSE rather than search within *effect* for *cause. When this transformation takes place in man's thinking science will have leaped ahead one thousand years in that day.*

VI

We again say these words were not written as a text book for the unfolding of new worlds of knowledge which lie ahead, unless such knowledge is distinctly relative to the one purpose for which it is written. We feel, however, that a few paragraphs of explanation might help open the doors to new light for the researchers who are now finding a "somethingness" in that supposedly "nothingness" from which "matter emerges only to be swallowed up by it."

What researchers are now discovering is that there are two ends of the creating units of matter. One end is the zero fulcrum end which projects dimensionless Mind-imagined patterns in pairs of bodies, and the other end is the enlarged pattern in motion. Their relation to each other is in the shape of a cone.

The "anti-matter" end is the apex, and the projected "pro-matter" end is the base of the cone. To restate it in electrical terms, the apex end is the cathode and the base end is the anode pair. As the anode pair always move as separate units of a pair, until they become united as one, so, likewise, do their apice projecting fulcrums *seem* to move to control their balance in space. *Naturally the zero universe is motionless, and no movement takes place at the zero fulcrum, for movement does not begin until division of electric thinking projects a two-way wave-lever from the fulcrum to manifest its energy and its Idea.* It is the wave-lever which moves, however, and not the fulcrum, and when balance moves in two directions it is always centered by one, no matter how it moves. See figures 17 to 29. This is too long a story to amplify here, but it is sufficiently relevant to our purpose to print seven diagrams here which will begin the clarification of the method the Creator uses in thus projecting His imaginings into the forms of His imagining upon the three-dimensional screen of space, which seems so real to us. The day will come in man's unfolding Intelligence and innersensory perception when he can clearly see that there is no real material universe, nor is there dimension. *There is an illusion, however, which stands for the reality of the substance it simulates, and that simulation is interpreted by our senses as motion. When man can comprehend that he will then know that what he thinks of as motion is, in itself, but simulation, for motion, likewise, is but illusion.*

The seven diagrams will also help give more meaning to the new idea of anti-matter, which has just entered the scientific world, as well as to open the door to the coming of the new science of Space Geometry.

In Fig. 62 you will note a centering point marked zero. That is the beginning point in the Mind-universe which marks the eternal Identity of one unit of a multiple repeated Idea. Your Identity begins and ends there. Likewise, the Identity of every creating thing in the universe begins and ends there. It emerges from that point millions of times, and returns to it millions of times complexed by union with other Identities, which become a part of its purpose. That zero point is the Soul-seed of the oak,

Matter begins by division of stillness in the inert gases and its projection by two-way motion into pairs of octave wave tones, which we call elements. The principle of projection is the same that man uses in projecting the concept of his idea into the moving form of the idea. It is the dual radar principle of Nature. Reduced to its ultimate simplicity, Nature is a series of echoes.

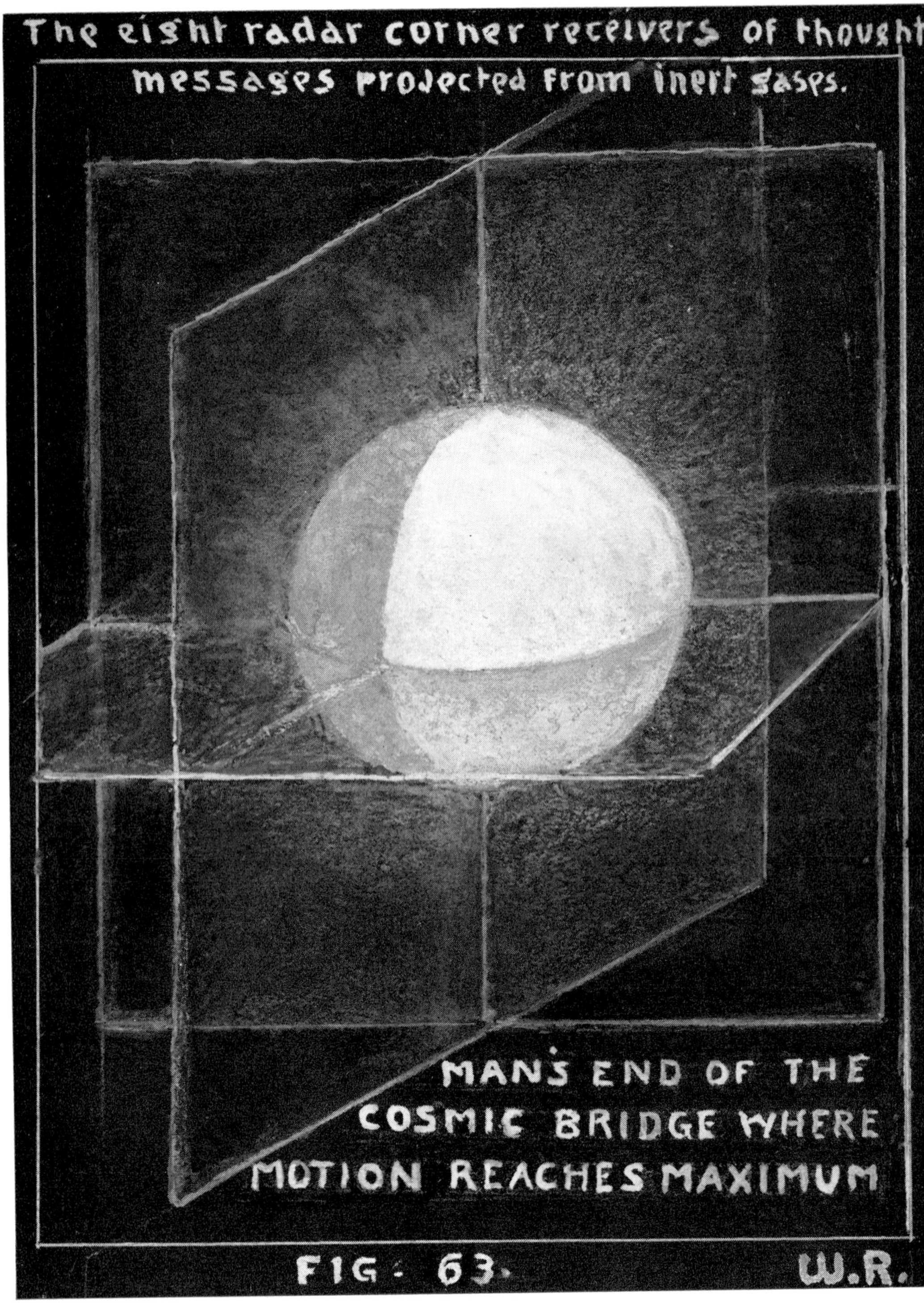

The cube projector, shown in Fig. 62, is repeated in the cube receiver, shown herein. Motion which has been divided into pairs and thrown out into space by the projector, is multiplied in the vortices of the receiver, which corresponds to the corner reflectors of radar. Forms and sounds which are born in space are thus echoed back to space.

Every form of matter which is projected into space has its anti-matter image within space, which ultimately voids it and reborns it. Anti-matter concepts have their own balancing relation to their material images in space geometry.

or of a corpuscle in your blood. You, and it, and all things else, are the formed-images which simulate the Creator's imagining. You, and the oak, and all things else, are thinking God's imagining. That zero point is *you*. In it is all-knowledge and all-power to extend to as you desire it, and are aware of it within that "kingdom of heaven" within you. You must some day learn that. You must in time learn that the eternal YOU is on the other end of your projected body. If, at present, you are limited to body-awareness, it is only because you do not know of your divine centering Source. Many people in the world have hardly any awareness of the God which centers them. Science is at last beginning to realize that there is an anti-proton, and an anti-meson and is searching for anti-other things. That is the most wonderful thing that has happened to the intellectual world in thousands of years. *It has not yet become aware, however, that the anti-matter universe is God's universe.*

Fig. 62 is the seed. When it is ready to extend—or to be reborn as an oak, or man, or carbon atom, it divides and becomes the fulcrum of itself. The Mind-projection mirrors then provide a cube wave-field for projecting idea into measured form of idea. By studying figures 62–63—and 64—you will also see why nature can never pass beyond the sphere in form. That is the end of its journey. The reversal of polarization begins there. The charge then becomes discharge. Spheres then oblate by throwing off rings. In ordinary language life is maximum there and death must take over. Life and death are born in the same cradle but they meet at that point as equals. That is the basis of the radar principle. The end of the journey of sound, as of all things else, is in one of the eight corners of the cube wave-field. Sound must return from that focal point. It is "reflected" from there. Electrical and radar engineers recognize that fact. They have even coined the name of "corner reflector" for it.

In Fig. 62 you will see eight sections of a cube wave-field, which constitute the four pairs in every octave. Each of these pairs is a radar corner reflector in reverse. The reverse is a corner *projector*. The corner *reflector* of radar is a mirrored image of the corner *projector*. The four pairs constitute a cube wave-field. The cube projector never appears with the seed until it "germinates," which means that the conditions for its rebirth are normal for that rebirth. If it never again germinates on this earth it still is and maintains its one plane existence eternally. It cannot be destroyed for it is Mind-Idea of Mind-thinking, and that is what constitutes all Nature.

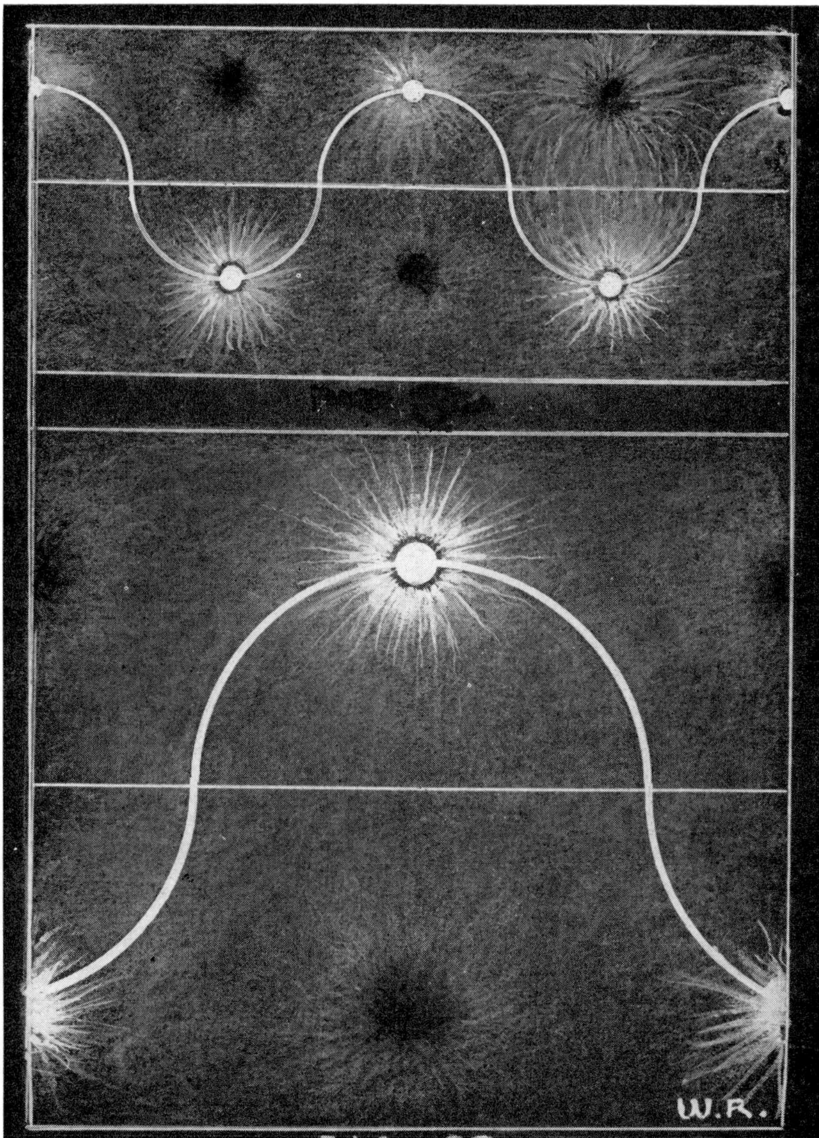

FIG. 65

The pendulum which swings one way has its invisible counterpart which swings the other way. This drawing illustrates a sun at the crest of its wave which will eventually be voided by its counterpart, awaiting at its trough, and reborn again from the same point. See Fig. 66.

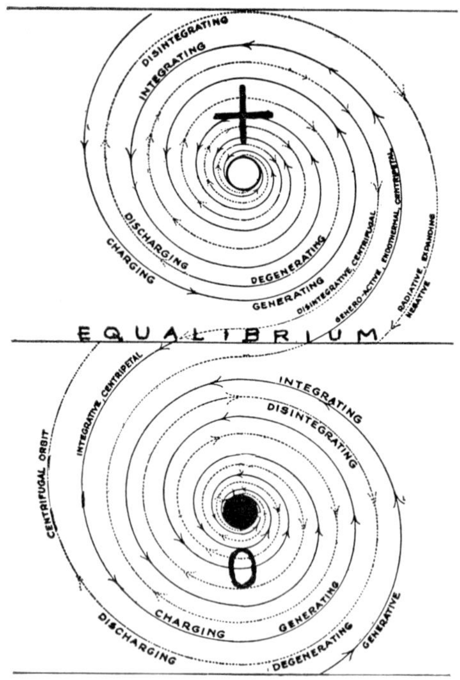

FIG. 66. Exemplifying the perpetual interrelationship of matter and anti-matter as shown in Fig. 65, which gives and regives to each other for perpetual sequences of rebirths into each other. The outstanding characteristic feature of Nature is repetition. Life and death are other names for it. Reincarnation is still another name for it.

Fig. 63 is the body end of the cosmic bridge. It is that which is projected. It is motion compressed into a spherical image of the transient form that is eternally recorded in the hole which is within the four rings of the seed. That sphere is also composed of eight parts, and is centered by the zero of its Source.

To more fully comprehend this omnipresent radar universe of reflecting light mirrors and lenses, take one of the eight reflectors out of a cube, or make one composed of three mirrors placed the way they are shown in the corner of Fig. 62. If you analyze

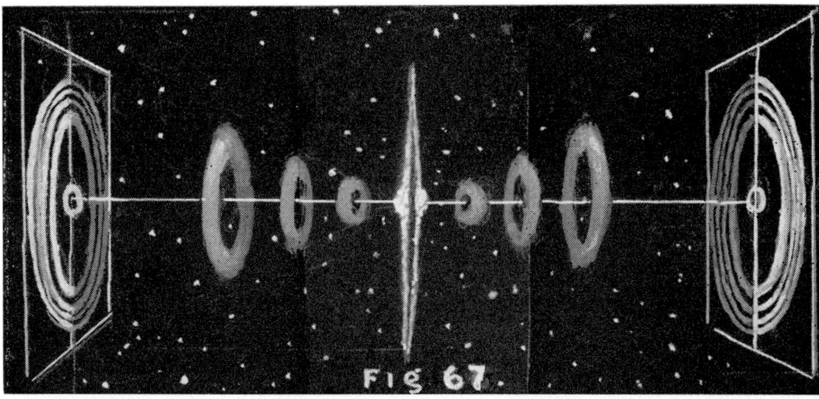

Illustrating principle of projection of Mind-desire energy into pairs of tones to create the nine octaves of tones, which constitute Nature. The spiral nebula—seen edgewise—which occupies the amplitude position in the wave current, corresponds with the loop of force which centers every polarized unit of an electric current.

this shape you will see that it is half a cube cut through its diagonals. If you now slowly insert an object on the end of a long needle, toward the apex of that pyramid you will see it reflected three times, and those reflections will grow ever closer as your object nears that apex. That illustrates the multiplication, or compressive power of Nature. It also symbolizes centripetal force and higher potential. It also represents three of the octave elements on the red or blue side of the whole octave, of which the three mates are in the reverse corner reflector.

The fourth of the octave pairs is the sphere itself. These four pairs, and their inert gas recorder, constitutes the whole octave wave-field principle. It shows how matter is constructed. The balancing dark images in figures 64 and 65 show how the zero universe balances every movement in the universe by centering every unit of matter with its accumulated potentials at a zero point within it, which we call gravity, and with another equal zero point at balanced focal points between each pair.

In Fig. 62 you see God's end of the bridge between the Creator and His Creation. In Fig. 63 you see man's end of it. In Fig. 64 you see how both ends are united by the electric wave, and in Fig. 67 by the electric current. In Fig. 67 a nebula has

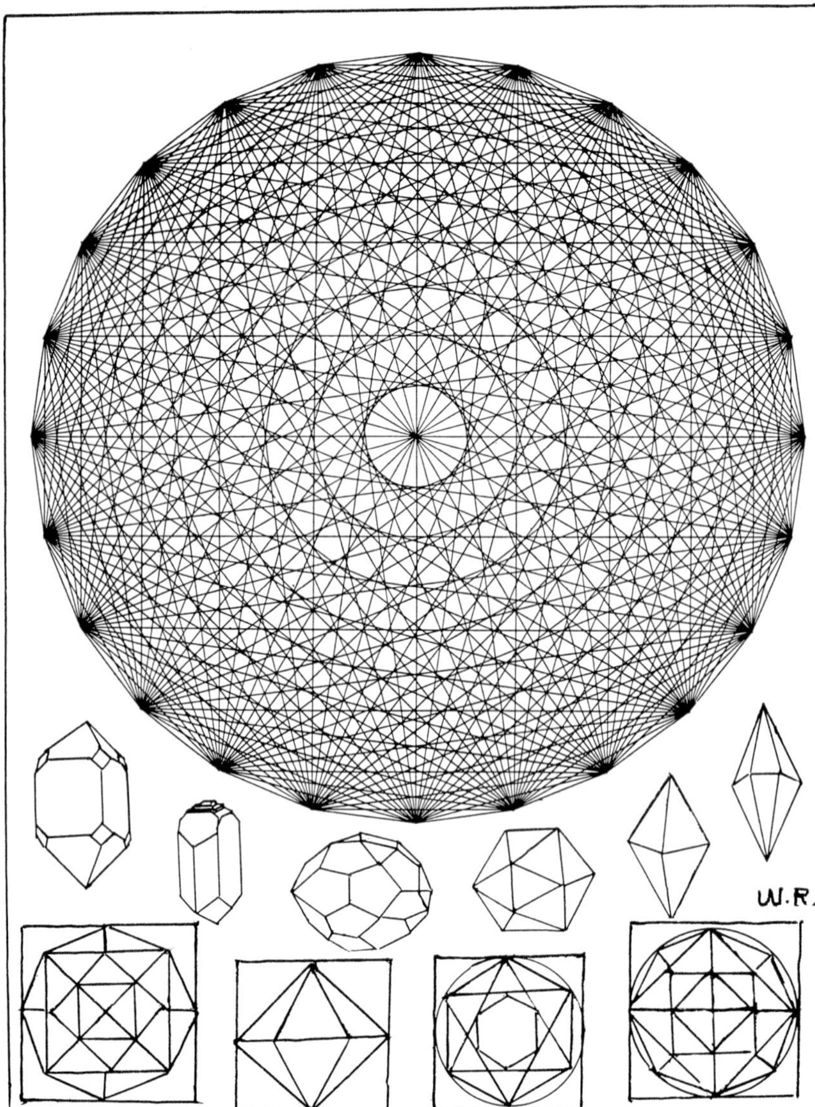

FIG. 68. Indicates a basis for the universal build-up from two-dimensional plane to three-dimensional cube, and back again to disappearance through two-dimensional plane.

been placed in the position in which a loop of force would occur in the electric current, and where a carbon atom would be found in the elements of matter.

These drawings should clearly demonstrate how the Intelligence of the invisible zero universe absolutely controls the visible universe of motion as dynamically as the operator in the cinema controls his visible idea in motion. This control of the visible universe by the invisible will become dynamically applicable in man's laboratories, and his creative efforts in discovery and invention, in the ratio in which he comprehends the geometry of space. That in itself, is an exhaustive study, a practically new science which could never be fitted into the old concept of matter, but can readily be fitted into a universe based upon nine octaves of mathematically rhythmic cube-based tones.

VII

The question has long been asked by research scientists why it is that the inert gases will not mix, or unite with "any of the other elements." The first answer is that the inert gases are not electrically divided and conditioned elements, as all of the others in the nine octaves are. The inert gases begin in the first octave as invisible white fluorescent light of zero motion. They end at the 9th cathode in the 9th octave, as visible white fluorescent light, which has reached a speed of nearly 186,400 miles per second. Fluorescent light is that light which begins in the undivided electric spectrum. It is the beginning and end of motion. All motion is either red or blue, according to its sex. The end of motion at the amplitude of the 9th octave means that the divided spectrum has been united as one colorless, sexless light which has been under such high compression that it has reached its limit of conditioning by motion and must be transformed from the white light of visible motion to the invisible white Light of Magnetic stillness. The fluorescent light is that ending of electric power to divide motion into pairs, and to condition the pairs with the opposing sex tensions of electrically divided spectrum opposites. The inert gases are not pairs. They are not divided. Division takes place by light projected from them, but that projected light of spectrum pairs is the basis of the electrochemical elements, which have great volume and density in comparison.

We have described the "bullets" of white light which are projected from radium. The speed and microscopic volume of the white rays of all of the transuranium elements acquire their maximum fluorescence and maximum speed of terrific compression as they withdraw into their inert gas niton, and from there on down the line to helium. On the return journey, however, this maximum fluorescence also acquires maximum penetration. It is this power of penetration into the 4th octave pressures of your body which expands its cells and brings death to you and all 4th octave organic bodies. It is also this power of penetration and speed of the 9th octave which not only unlocks the pressures of the lower inert gases, octave by octave, but also unlocks the pressures of the 8th, 7th, 6th and 5th octaves of radioactive elements. Most deadly among these are radium, thorium, actinium, strontium, barium, calcium, potassium, arsenic and phosphorus.

These are the elements which radioactivity has released in vast quantities in the upper stratosphere and would soon crowd the lower atmospheric layers with such a change of environment that oxygen-dependent bodies could no longer live. It may seem incredible to you who are accustomed to the idea of density in matter, and of the difficulty of ordinary light to penetrate even its crust, to know that the white fluorescent rays of plutonium or neptunium can penetrate several feet into solid lead after having become "de-electrolized" by their impact with niton. It is the accepted belief of the entire scientific world that radium eventually becomes harmless by "decaying" into lead. This is an impossibility of Nature for radium is male, and lead is female, in their sex divisions. Such a transition is one effect which is entirely outside of Nature.

As a final word white light is the end, as well as the beginning of the divided spectrum. Divided light reaches its limit of power to become compressed and from that point its power of expansion—or explosion—into its zero is indescribable by words. If such rays as these reach your body they can penetrate to any part of it without resistance. They can also act as triggers to unloose all other lower octaves which are stored in them. You thus have uranium in your blood cells and tissues, strontium and calcium in your bone marrow, and other radioactively liberated particles affecting every part of your body, its genes, hormones and even unto the destruction of the proteins of your hair and nails, all confined within one particle.

Fortunately, the human body can stand quite a dosage of such rays before showing it by bone cancer, leukemia and other effects of radioactivity, but also *unfortunately,* these rays are accumulative. *They never leave the body.* The defects mentioned above, including sterility, are incurable advance notices of the fact that you have reached your limit of power to take more of these translucent rays into your body. You can take no more, yet what are you going to do about it if you are irrevocably sealed within an increasingly radioactive environment? What is any living animal or vegetable going to do about it? *That is the question which we are submitting to the world. What is your answer to it?*

VIII

In conclusion we must add a few more paragraphs regarding the construction of the inert gases, which need volumes instead of paragraphs. This we must say, however.

If you refer to Fig. 5, which symbolizes the universe as a nine octave harp, you will note by the drawing that the strings are of the same length. It would be impossible to symbolize their relative lengths in trying to express their underlying principle. Likewise, it would be impossible to symbolize their ring proportions. All we can say in this brief treatise, to clarify that mystery, is that the four rings of the 1st octave are very large and grow smaller as the octaves proceed to the 9th. Likewise, the length of the 1st octave string is so very long that its vibrations are immeasurable and undetectable. The only thing we can think of to compare with this first string is the lowest string of a harp. It is so low that its vibrations have no resonance in them, but they are detectable. Now add one or two much longer and thicker strings below this lowest one and they would be silent, for it would be impossible for them to create one common wave-crest amplitude where a collision could occur.

The reverse condition takes place in the very short strings of the 8th and 9th octaves. These are so tightly tensed by maximum compression that they too cannot reach one common amplitude. Even in the 6th and 7th octaves that attempt to collide and form a "loop of force" is split into five pairs of attempts in each of these two octaves following the silicon octave. In the 8th there are thirteen pairs, but in the 9th the number is unknown, and at least twenty-one on the red side. The blue side fails to provide balanced mates for the six of the transuranium series, so all motion disappears into zero at that 9th octave position.

You may be helped in imagining this condition by turning up the tuning key of the highest string on your harp to a point where it will no longer produce a normally advanced wave-crest tone. If the wire did not snap you would eventually reach a position where you could not hear any tone which would be audible to your senses. We trust this brief explanation will at least direct your thinking and imagining to the incredible high frequencies of such an element as plutonium, and the unbelievable high fission temperatures for infinitely short periods, which makes it possible for the radioactive elements to prepare humus and alluvial soil for organic life, by their terrific explosive power on micro-scale. We trust also that this description will fully explain to you why such extraordinary power for expansion should not be released within the cells of bodies which are struggling hard to keep their body cells compressed sufficiently to live. If you could imagine just one cell of your body struggling against one radioactive ray which is expanding it, the comparison would be as a mouse straining against the power of an elephant. You will admit that such a discrepancy of balanced power would be hopeless—and it is.

CONCLUSION

We now conclude our story, which has primarily dealt with the so-called life and death of bodies, for they are what we are interested in preserving from destruction in order that they may fulfill their purposes of completing the Divine Symphony of Creation as actors in it. To do so comprehensively, however, we have had to tell enough of man's divinity as an eternal Being to enable one to gain a better understanding of his universe, and the Creator of it.

We, therefore, repeat words written in earlier pages of this book, which say that our bodies express life only by the fast generoactive centripetal motion which charges and polarizes, and that they die only by the fast radioactive centrifugal motion, which discharges and depolarizes. We made this statement to demonstrate that radioactivity multiplies its death speed in each higher octave. In so doing we demonstrated that a radioactively charged atmospheric environment would sooner or later eliminate all organic matter from the surface of the planet and reduce it to desert of rock and sand, broiling in a temperature of at least 180 degrees.

This was our primary purpose. Its secondary purpose was to hasten the day of spiritually unfolding man by giving him new knowledge of his close relationship to His Creator and to Creation. This new knowledge of the Mind-universe, in its relation to the body-universe, would also help clarify the nature of radioactivity in its action upon organic bodies, and especially oxygen. As oxygen cannot co-exist with the free metals above the carbon octave upon the red side of the spectrum, radioactivity must not be allowed to change our atmospheric normality.

This information has been given for you to act upon in accord with your decision and belief. If you believe that what is herein written is not true to the facts of Nature, the only other alternative is to prove that we are wrong by continuing the practice and pay the dreadful price of such a gamble, which these words were written to save you from. It would be a very pitiful thing to happen, however, for the human race is now due for a tremendous mental transition into another step of higher intellectual unfolding. If another annihilating third world war can be avoided, our future generations of gradually unfolding Mind-awareness will bring to man the knowledge of his own divinity.

As long as man continues to think of his *body* as being his individuality he will think individually, rather than universally. When he, at long last, discovers his eternal Self to be the projector of his own body, and is making himself in his own image, he will then know his universality, his Oneness with all Creation. He can never thereafter think of himself as an individual and will so act in all of his transactions in relation to his Oneness with all other men.

This ideal higher condition will come about more quickly as the new world's knowledge of the geometry of space, and the Oneness of all creating things, shall unfold to make the zero invisible universe as intellectually real as the universe of motion is now sensually real. This day will come, whether civilization again falls or not. A few centuries of dark ages could not stop it, for physical man would still be here. *Radioactivity alone could stop it, for it would take at least 50,000 years for the recovery of the earth's normalcy,* and millions of years after that for the slow unfolding of the pattern of the man-idea through the many body-building stages, which Nature has to make use of in her long-time processes of unfolding man of today from the amœba of millions of years ago.

Civilizations of individuals may come and go, as many have come and gone, and as this one is now so rapidly declining, but always the *cause* of decline is individual thinking and of individual concern. Individuals believe they can acquire by *taking*. Universalists know they can acquire only by *giving*. When all men shall have learned the lesson of love sufficiently to become ONE in their thinking, they will become one in their giving to the WHOLE instead of to the self. When that day of unity shall come we shall then have a civilization which will endure, but not until then.

OMEGA

The supreme miracle that can happen to any man is the discovery of his own divinity in his Oneness with God. If this book shall have led him to the Light of Mind, which man is in God, and he is illumined with that Light of all-knowing, all-power and all-presence, giving him dominion over all the earth—it will have then served the double purpose of the hopes of its authors—rather than just the single purpose of its naming.

ADDENDUM

Non-Conformity of Lee-Yang Theory

As THIS BOOK goes to press an experiment at Columbia University conducted upon theories promulgated by two young Chinese physicists—Professor T. D. Lee and Professor C. N. Yang—affords an excellent example of how easily the senses can be deceived into arriving at wrong conclusions. We are, therefore, adding an addendum to explain why neither the premises, nor conclusions, are in accord with Nature.

The essence of what made this discovery so revolutionary that it upset many fundamentally believed in facts of physics lies in the conclusion that all atoms spin in the direction of their motion (meaning the direction of their axis), which means that atoms or particles ejected from the north end of a "decaying" atom, such as cobalt, spin in a right-handed (clockwise) direction, and those which are emitted from the south end spin in a left-handed (anti-clockwise) direction. This discovery led to the possibility that atoms acted like right and left-threaded screws, and that might eventually explain how matter is held together by some mysterious force *within* matter and solve the mystery of Creation.

This cannot be briefly explained because every single premise upon which the experiment was founded is wrongly assumed to be what it is believed to be. These wrong conceptions must first be clarified. We will then explain why cobalt has given an entirely different result than carbon would give, and that sodium would have given another effect of unbalanced ejections, while sodium chloride would give a balanced effect, and tantalum would have given another effect from the one produced by cobalt. In other words, when it is shown WHY cobalt emitted a large electron from one end and a small one from the other, and it can be clearly explained WHY every single unit atom and every paired one will give different, and predictable results, it will then be easy to comprehend why the conclusions arrived at by this experiment are invalid. To begin with, the experiment itself should not have been necessary. With a true knowledge of the Nature of electricity the right results would have been predictable. To substantiate this statement we will predict other results from the same experiment with full knowledge that they will be as stated.

STEP BY STEP EXPLANATION

Step No. 1. We quote from "Life:" *"A crystal containing cobalt nuclei was placed in a magnetic field created by current flowing through a wire coil. It was already known that the cobalt nuclei would line up in the magnetic field."*

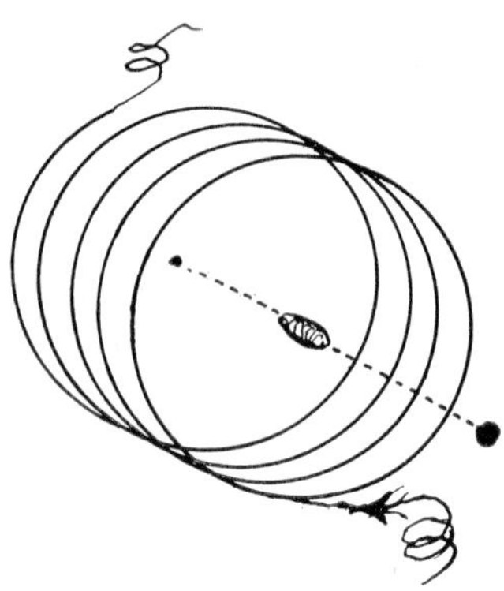

FIG. 69. This is the way man divides stillness to produce motion.

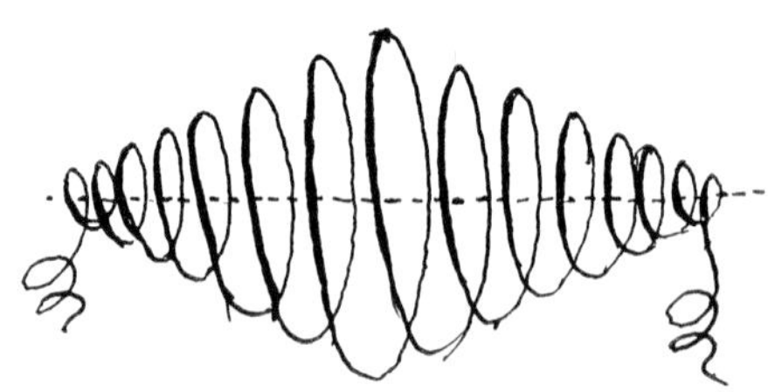

FIG. 70. This is the way Nature divides stillness to produce motion.

Fig. 69 indicates the wire coil, the direction of the electric current, and the position within it that the cobalt crystal is placed. The first wrongly assumed premise is that the wire coil creates a magnetic field. It creates an *electric* field, not a *magnetic* one. There are no magnetic fields in Nature. There are electric fields of curved motion, which are bounded by magnetic planes of non-motion and zero curvature. It must be borne in mind always that any expression of motion anywhere, is omnipresently repeated *everywhere.* The human consciousness has not quite acquired that sense of omnipresent repetition as yet but familiarity with it through TV and radar is developing it. This cobalt experiment is, therefore, not a local happening which is taking place in Columbia University and nowhere else. The electric field provides for it to happen there but the magnetic geometry of the zero universe provides for it happening universally by Nature's radar system of extensions from zero points of stillness, which become mirror planes of stillness invisibly formed in space to repeat the divisions and multiplications of motion in any part of the universe to every part of it.

The wire coil of the Columbia experiment has given no thought to the basic fact that electricity divides for the purpose of multiplying. Otherwise, the coil it made for its electric field would be like Fig. 70.

We cite this fact for any experimental work done in a laboratory is valid only if it conforms with the electric current of Nature, which alone creates the universe. Any experiment, whatsoever, which does not fit *within* the principles of the electric current is invalid, for nothing *outside* of Nature can be valid.

Step No. 2. The cylindrical coil made for the creation of an electric field does not, in any respect, conform with the electric current, which divides the stillness of cathodes into moving pairs, for the purpose of multiplying the motion of those moving pairs to create the heat and compression necessary to simulate the IDEA, which concentrated electric Mind-thinking desires to give objective expression to. The Einstein Equasion properly defined this principle by the words which say that fast motion multiplies electric potential, but that principle has not been applied to the mechanics of coil making. For that reason the loops of force which cylindrical coils create, are not focussed as they would be if they followed the crystalline shape of the balanced electric current as shown in Fig. 70.

If you will study the corner reflectors of radar, as shown in figures 62–63, and their relation to ring projections from inert gases, and observe their spiral formation, as shown in Fig. 64, you will better comprehend the wave, which controls the construction of all matter.

SUMMARY OF STEPS—NUMBERS 1 and 2.

The two essentials omitted for the perfect carrying out of this experiment, up to this point, are 1—failure to conform to the working principles of the electric current, and—2—failure to conform to the omnipresence of Nature by making electrical fields to conform with the mathematics of dividing and multiplying pressures.

Step No. 3. We again quote from "Life": *"The test was based on the previously proven fact that the nucleus of the atom 'decays.' In the process it emits one electron and one electrino (an uncharged particle even smaller than an electron)."*

It was not necessary to prove that cobalt "decays." To decay means to disintegrate, depolarize, discharge and die. All matter does that, whether atom or man. This experiment, however, did not assist in the decay process, for it lined up the gravity shaft of cobalt with the gravity shaft of the electric field. *That is a process of polarization, which divides any one electrical condition into pairs.* The process of decay is an opposite effect. It ejects rings from equators, as clearly shown in Saturn—see Fig. 4—and in Jupiter—see Fig. 3.

Also the statement that uncharged particles are ejected from one end is not true to Nature. There are no uncharged particles in Nature. Wherever there is matter there is motion. Motion is an electric effect, which charges all matter.

Step No. 4. *"According to parity, as they spun half of them should have emitted their electrons from the "north" end of the magnetic field and the other half toward the "south." Actually far more electrons came out of the "south" end of the nuclei. By repeating the test under varying conditions, it was verified that cobalt nuclei emit electrons from one particular end. This shows that one end of the nucleus is different from the other, something never known before."* ("Life")

Let us analyze these statements to see if there is right reasoning in them in accordance with the electric current. In the

"Life" diagram, large particles, labelled electrons, are being ejected from the south pole of the cobalt atom and small ones, called electrinos, are being ejected from the north end. Fig. 69 is an interpretation of the "Life" diagram excepting that we are using only one particle. The "Life" diagram shows three cobalt particles, oval shaped, and their respective small and large ejected particles. The names of these particles is of no import here, but the fact that one end of cobalt atoms ejects big particles, and the other ejects small particles, is important. Also, the statement that it was never known before that one end of an atom is different from the other end is also important for it is not true. It has been known as long as electricity has been known that each end of a polarized body is different, and all matter is polarized. Each end of a balanced atom—which means equally paired—will emit the same sized particles from each end, because polarized, or divided pairs, are male and female and belong to the red and blue sides of the spectrum. *The important fact that has been overlooked is the fact that an unbalanced particle of matter cannot emit equal potentials from both ends.*

To demonstrate this fact put a cobalt, or steel needle in the position given to the cobalt particle. *You will see that both ends are different, and are not interchangeable, unless repolarized in reverse.* You might better comprehend this if you lower two needles toward the center of the coil. You will find one reaching out for the north and the other reaching out for the south. Now try again with a steel or cobalt rod as long as the field. By placing it in the conical coil shown in Fig. 70, you will now see that not only are both ends different, but also the middle is different. Both ends are polarized gravity centers for high electric potential, and the middle gives no evidence of gravity, for the electric potential is zero in the middle. (In an unbalanced metal, like cobalt, the middle is a little off-center.) *The reason for that is that the middle is the cathode dividing point, and the ends are anodes, which unite as one.* In other words, the reason both ends are different is because one end is red—or male end—and the other is blue—or female end. Cobalt is preponderantly blue, or female. The red end of the spectrum is weak in it. Under such conditions of unbalance the polarization process could not possibly eject equal potentials.

To better understand this take the cobalt out entirely. Only the zero of equilibrium is now there. Now turn on the current and you immediately divide that equilibrium, balanced zero into four opposite and equal pairs, all of which units are in perfect balance with their mates. The cathode is also in the exact middle instead of being off-center. *That is the way Nature creates atoms.* Cobalt is an unbalanced unit of a pair. It cannot be divided into equal pairs when used as a cathode.

If we now put a piece of the metal cobalt *between two coils* like Fig. 70, the middle will become the plane of highest potential and its center will "attract" a needle if held toward any point of its circumference. If cobalt is thus placed in the anode position, instead of in the cathode, there will be no ejection of electrons and neutrinos from its polar ends. Instead of that, there will be ring ejections from the equatorial plane of maximum high potential, maximum heat and maximum density. We now have the decay process instead of the polarization process.

Step No. 5. One should now be able to fully comprehend that if one attempts to divide an unbalanced element like cobalt, by polarizing it, one will obtain an unbalanced result. Carbon, for example, is a balanced dual element. It is true cube and is white, while cobalt is distorted cube and is blue. The same experiment tried with carbon, or sodium-chloride, or potassium-bromide, would give balanced results. The ejections from both ends would be the same. If, on the contrary, sodium-iodide, or potassium-chloride is used the results would be as unbalanced as these latter pairs are unbalanced. Each of these pairs crystallize in distorted cubes while the evenly mated pairs crystallize in true cube, like carbon.

The familiar bar magnet, created by the electric current, will serve as a clarifying example of balance in the elements, which are all created in pairs, but are as often united as unbalanced pairs as balanced ones.

Step No. 6. In further statements made in "Life," "SCIENCE NEWS LETTER," and many other newspapers, it is claimed that the discovery of different ends to atomic particles, and emissions of other particles from opposite ends, gives to each emitted particle *"a spin in the direction of its motion."* As the motion of these ejected particles is claimed to be in opposite directions the result of the discovery has "proved" that motion is two-way, *"which means right-handed and left-handed."*

To explain to you how utterly impossible it is for this to take place in Nature, and again to show how easily the senses can be persuaded to believe as facts the very opposite of what is factual, we have to go back to the utter simplicity of the fundamentals of our universe and lead up to a comprehension of such an impossibility. We will do this as simply and briefly as possible. In other words we will return to the electric current and compare this experiment to just one cycle of it. *We, therefore, first postulate the basic fact that, in all this universe, there is only one direction to motion, but that one direction produces two opposite series of effects.*

Rain falls to the ground and rises as vapor. That seems to be two directions. A cyclone is anti-clockwise in the northern hemisphere and clockwise in the southern hemisphere. Right-handedness and left-handedness, up and down, in and out, hot and cold, life and death, expansion and contraction, and a hundred other opposite pairs of effects are produced by one direction of motion in this pulsing universe. If we had to clarify this idea for a teenaged boy we would tell him to take a pencil and a large piece of paper, then begin drawing a closing spiral, clockwise, starting slowly from way out into space, and increasing his speed constantly as he turns around a hole in the center. We would tell him to keep on drawing in the same direction around that hole but go outward spirally and decrease his speed continually until he got way out into blank space again. We would tell him to keep on repeating that operation over and over again, without changing direction.

Let us now look at this drawing, Fig. 71. Two effects have been produced in this opening-closing universe of compression-expansion—life-death—and inbreathing-outbreathing sequences. Density has been created in the direction of the center and potential has lowered in the direction of the circumferences. So, also, has heat, and speed, and high potential been created at the center and their opposites out in space. Fig. 71 is the demonstration of this fact as drawn by a child for simple understanding.

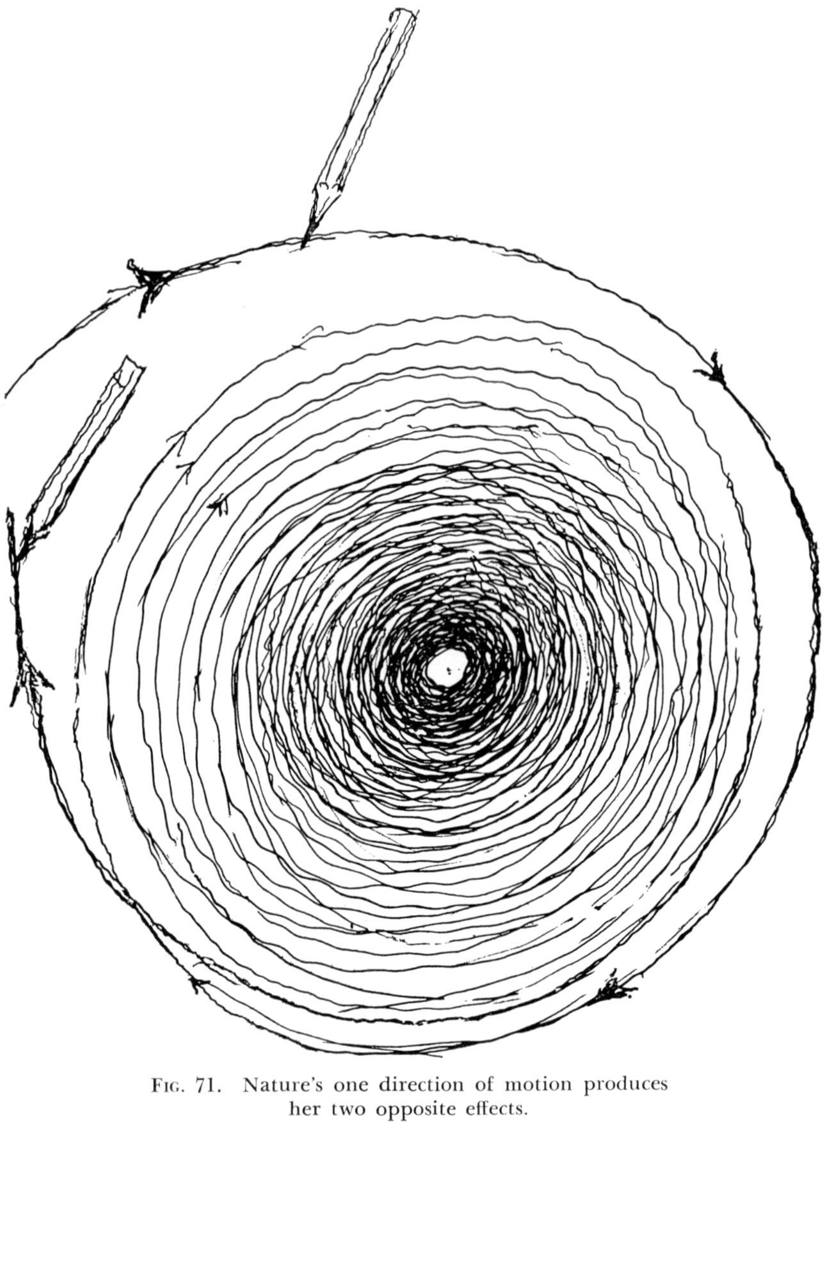

FIG. 71. Nature's one direction of motion produces her two opposite effects.

Fig. 72 demonstrates this one direction for a two-effect principle in small scale matter. Figures 73 and 74 demonstrate the same principle in large scale matter. Nature does not have one law for atomic scale, and another for stellar scale. She has only one way of constructing matter, and that is fully covered in the electric current and the octave wave.

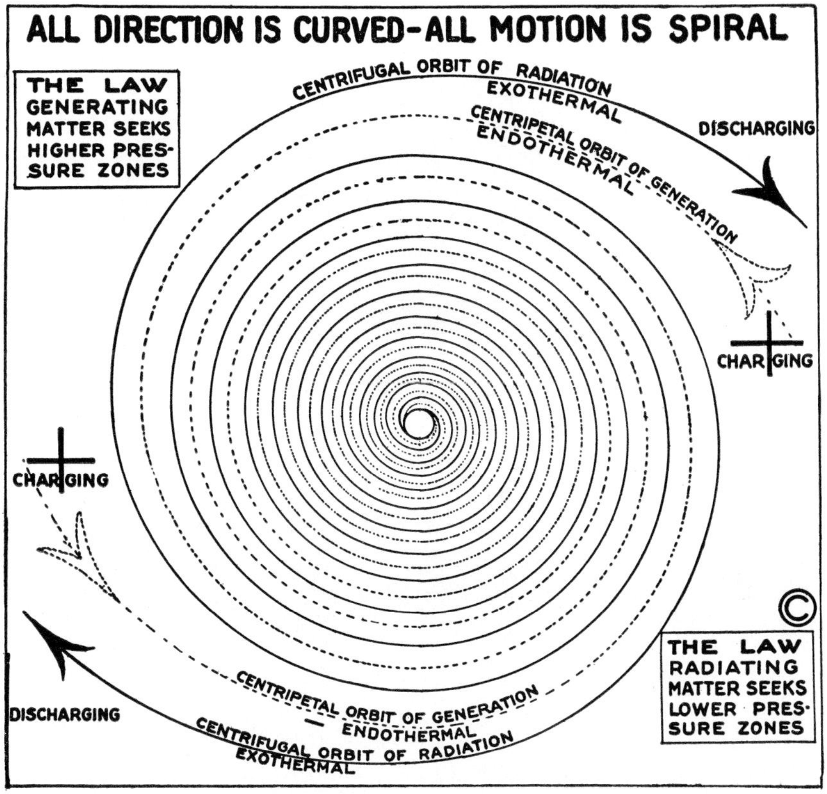

FIG. 72. This drawing exemplifies the electric effect of two seemingly opposite directions and opposite pressures from one direction of motion, as demonstrated in Fig. 71.

Step No. 7. *There is no right-handedness nor left-handedness in Nature, but its every effect causes the illusion of right-handedness and left-handedness. That illusion belongs to the senses, therefore, and not to Nature.* Our senses have become so fixedly

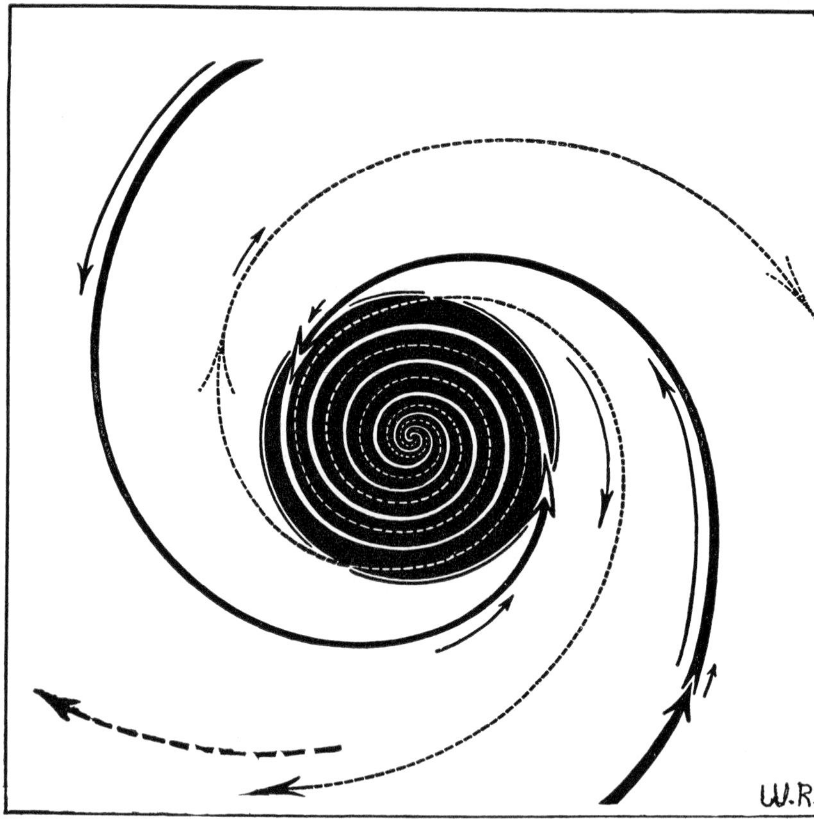

FIG. 73. This same principle, as demonstrated in Figs. 71–72 is also demonstrated by the atomic and nebular systems, which form where two anodes are united by collision. This figure demonstrates the one-way of motion which produces both effects of life and death in Nature.

accustomed to dual direction that it is difficult to explain it away, for the human understanding of omnipresent Oneness has not sufficiently unfolded as yet to make it easily comprehensible. We will endeavor to do so, however, by the use of child primer simplicity.

As a first step we again remind you that Nature creates everything in pairs, and that all pairs unite to form masses. The hemispheres of a sun are united pairs, for example. Try, therefore, as a first mental exercise, to image the uniting of two

FIG. 74. A familiar spiral nebula in the heavens, seen in the direction of its poles instead of the direction of its equator, as in Fig. 67. It has always been believed that two directions of motion have been responsible for this effect.

hemispheres of the sun, which are spinning in opposite directions. The fact that you, yourself, are right-handed and left-handed, adds to the difficulty of the subtle thing which we must explain about our universe, which has led these observers to form such conclusions. What we mean by this is that Nature is bilateral. Everything which is created is equally extended in pairs from a cathode dividing equator. The equator acts as though it were a mirror. It reflects each extended unit of each pair in reverse. These divided pairs unite as two hemispheres of a sphere, which is also divided by an equator.

As a second mental exercise consider your own body in this light. The equator of your body is a plane extending through it between your eyes, which divides your body into right-hand and left-hand units of pairs. *During the millions of years of your unfolding your senses have translated unitary facts as dual effects.* Let us demonstrate this idea. Imagine that you can reach out with your right hand and grasp the right-hand end of the earth's pole of rotation. With your left hand you reach out and grasp the left end of that pole. If you stand there, as rigidly as steel, the equator of the planet and the equator of your body are in the same plane. This being true one of your eyes is on the right side of the equator and the other is on the left. In this position both of your eyes see only the one effect of the planet's spinning in the same direction. Now bend your body so that both of your eyes can see what the right side of the planet is doing. You now see a clockwise direction to the spinning of the planet. Now bend your body so that both of your eyes can see what the left side is doing. You now see an anti-clockwise spinning. In other words you see opposite directions of motion where there is but one direction.

We will again demonstrate this by having you hold a pencil in each hand in such a manner that both touch the spinning planet at points to the right and to the left of its equator. As the earth spins you move both pencils slowly toward the stillness of the pole of rotation. If it were now possible to remove these two pencilings from the planet and lay them flat before you, you would again see them as oppositely directioned, right-hand and left-hand, clockwise and anticlockwise spirals, as shown in Fig. 66. *Your senses have told you that, because your senses, and your body, are dual.* Mind-knowing is not divided, however, so if you unite that divided pair of drawings you can no longer sense two directions. Try it. Paste the two penciled diagrams together. Hold them to the light and you will see that both opposite directions are one. If you fold Fig. 66 at the horizontal line, you would find that both opposite spirals follow the same direction.

This is exactly what the Lee-Yang experimentors have done. Their coil of motion, which caused the division, turns but one way, just as the sphere which you have held between your hands, turns but one way. The particles have been ejected from that coil in an angle of 90 degrees from their plane of motion yet it is now held that these particles turn right-handed from one end to turn in the direction of its motion, and particles ejected from the other end turn left-handed for the same reason. *This creates the utterly impossible condition of giving matter two shafts of gravity, which are at right angles with each other,* as shown in Fig. 75.

Nowhere in Nature is there a precedent for projected masses spinning in the direction of their motion. A bullet from a rifle spins around its axis of direction. Rifle barrels are spirally grooved to accentuate this spin. Even if bullets are spherical, like cannon balls, they would still spin around their axis of direction, for their equators must necessarily be at an angle of 90 degrees from their gravity shafts, and no mass in Nature spins in any other direction than that of its equator.

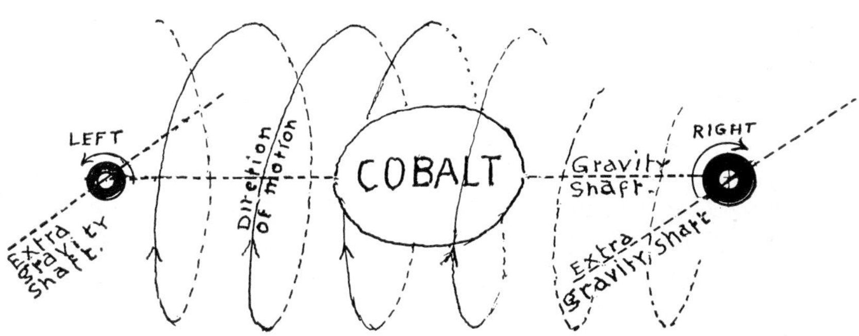

FIG. 75. A right-hand and left-hand spin in the direction of the oppositely projected particles, would require two extra shafts of gravity, at 90 degree angles from the gravity shaft of polarity. This is a violation of Nature's gyroscopic principle.

It is a commonly known fact by every physicist and laboratory worker that electricity runs only in the direction of the turnings in a solenoid coil. Every such worker is also perfectly familiar with the effect of polarization which evidences its effect at 90 degrees from the direction—and plane—of the electric current. These two seeming directions, and effects, are confusing to him for the idea of polarization, as an extension of stillness, has not yet entered his thinking. *It is new thinking but it must some day be realized that a three dimensional universe is only an illusion caused by polarization, and that there is no motion, whatsoever, in the direction of gravity shafts, which contribute to the three-dimensional illusion.* This leads us to the final step.

Step No. 8 is the last of the wrongly assumed premises which we must explain away. This last premise states, in essence, that as matter spins in the direction of its motion, and as there are two opposite directions of motion in which particles are projected from opposite cobalt ends, is it henceforth to be a law that: *"Particles will now be referred to as 'right-handed' or 'left-handed' contrary to what the principle of parity held."?* ("Life")

This is again difficult to explain in this early stage of Mind-unfolding. To attempt to explain it requires combating what our senses sense, and what our Mind knows. Again we will attempt it by child primer simplicity.

The fallacy of this entire statement lies in the belief that the two opposite directions, which the projected particles are following, *are directions of motion.* Herein lies the *cause* of the misconception, which pure illusion of motion is accountable for. These two opposite directions in which the cobalt particles have been projected are not the direction of motion, they are points or shafts *within the stillness* of the zero universe which have become identifiable by motion. The one direction of motion is that of rings which encircle that projected direction within stillness in planes of ninety degrees from it.

The human Mind has not quite progressed to that point where it can separate illusion from its cause, and our difficulty in clarifying the illusion is due to the aeons in which we have seen boats, planes, clouds, people and countless other expressions of energy move from place to place within the omnipresence of gravity to simulate that energy, which they are but expressing. Our senses have accustomed us to believe that the direction of motion is the direction in which we see an object change its position. Our knowing Mind, however, attributes that seeming change of position to gravity division of omnipresent Oneness, and attributes its cause to the direction of motion, which spins around this direction at an angle of ninety degrees from it. To clarify this new thinking we shall go back to *the electric wave within which all of the secrets of Creation are hidden.* Bear in Mind that the electric wave is the divider of ONE into four pairs which constitute ONE. As we explain this thought turn back to figures 17 to 29—to give objectivity to them.

Now vision the universe as all cathode—which means all zero —as visualized in Fig. 6. All motion begins at a cathode point in space, *which is all cathode.* It cannot move out of that omnipresent position, however, without dividing its seeming energy into four *radar-acting* pairs in the wave, which is beginning to construct, nor can it do so without measuring out its own *desire-energy* into countless billions of other cathodes throughout all omnipresence.

At this point turn back to the two cyclone drawings, figures 51 and 52. You will see there a zero which centers the cathode of that hurricane, which we call the eye. There is no object there, such as a mouse, or man, or thousand ton boat, or a crystal of cobalt, but there could be. It would make no difference as long as they are still and have not expressed a desire to move the whole universe, which must automatically move when they move. *Electricity divides stillness. Motion then begins, but not until.* It seems platitudinous for us to say that when we start a boat from that point, or a bullet, that neither of them are moving in the direction that our senses tell us they are moving, and it seems quite ridiculous to go beyond that and say that neither of them are moving at all in any direction, yet it is necessary to utterly change our knowing from what our senses tell us in this respect if we desire to comprehend the illusions of our three-dimensional mirrored universe and its very deceptive ways.

When our senses tell us that a speeding bullet is not a direction of motion it is difficult to think of it as stillness which is centering fast moving motion. It might help to remould your concept by imagining a motionless, floating cloud which changes its position because of a wind. The direction of motion is not in the cloud.

It is in the spiralling of the wind. The cloud is still in the same position in relation to gravity even though its position has been changed by motion, which is outside of itself, for it is still floating at rest. Even the slightest change in its position of stillness causes a spiral motion to encircle it and spin around it at an angle of 90 degrees from its gravity shaft, which extends from the earth. It may be quite an effort to comprehend that stillness cannot change, but can seem to. When you have fully grasped this fact, however, much that is complex will become simple. If these sincere young physicists had that Mind-awareness they would not have reasoned as they did in relation to an experiment, which they have mistaken to be something other than *the simple, basic polarization process of Nature which extends the two-dimensional thought-ring universe into a three-dimensional, dual action universe.*

Nor would they have come to the conclusion stated by them had they observed the majestic example in the heavens of all the planets, suns, spiral nebulae and galaxies which, without exception, turn in planes which parallel their equators, and always at 90 degree angles from their polar shafts.

Every planet and sun in the heavens spins in the same direction that a speeding bullet spins. That direction is 90 degrees from the direction of the speeding bullet, which is always the eye of the cyclone its motion is producing.

Their particles are similar to speeding bullets, or equators of planets and suns which always spin in spiral rings around their gravity shafts, and never in the direction shown in their explanatory drawings.

To aid in making such new thinking clear we will endeavor to express our thoughts by using the language of diagrams and familiar examples, which we will confine within one cycle of the electric current in principle. In thus confining our description of this new thinking within the demonstrable effects within one cycle of the electric current we sincerely hope that future thinking and reasoning will have a better basis upon which to form conclusions, which are not affected by the illusions of motion, that have for so long, and so disastrously, affected human reasoning.

There is, in this universe, an underlying principle which has always eluded mankind. If we try to put this principle into words they would read as follows: *Every effect of motion is voided as it occurs, is repeated as it is voided, and is recorded as it is repeated.* Our *senses* deceive us in this respect. They see the pendulum swing one way, but the Mind knows that an equal force simultaneously swings the invisible one the other way. That which is divided on one side of the cathode is equally divided on the other side. Likewise, that which is multiplied as it approaches the anode on the red side multiplies equally on the other side. Even time has its backward flow to make time stand still on its zero mark. The pendulum well illustrates this duality in Nature, which makes so many things seem to be other than what they are and to be moving when their moving is voided by an opposite moving. Nature is continually moving to the senses but within Mind-knowing it stands perfectly still as but a thought-picture of Idea. See Fig. 76.

You can readily see why it is difficult to find adequate words to explain a so seemingly universal paradox, and why we must use another language than words. This we shall do step by step very briefly—but sufficient to explain why both ends of the cobalt atom in the experiment emit different sized particles around a center.

1. We must know why this universe is bilateral instead of radial. This means that we should know why mass spins upon a shaft instead of around a center. This also means that we should know why the inward explosion, which basically causes motion in this zero universe, is a centripetal spiral, and why it is returned by radar to the zero of its source as a centrifugal spiral. If we first grasp this one idea we will then know why electricity explodes from the inside outward to divide stillness into pairs of spinning rings, and we shall also know why electricity compresses inward from the outside to form solids. Let this be our first lesson in CAUSE.

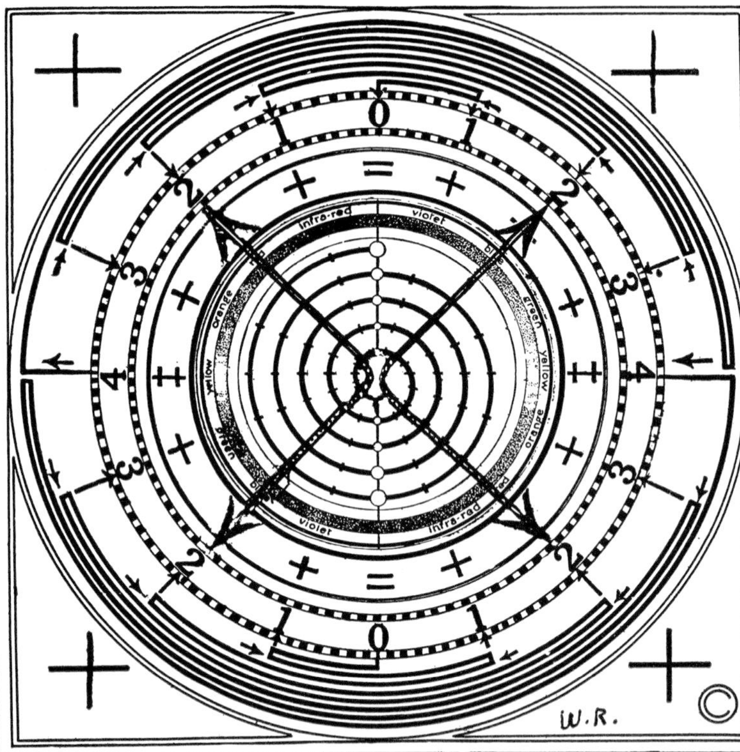

Fig. 76.

Every effect in Nature is divided into pairs of opposites. Each one of each pair is the reverse of the other. Each one is like a mirror reflecting the other. Nature is like unto a clock with two hands which bend away from each other in opposite directions but equal potentials. If one hand multiplies potential the other simultaneously multiplies it equally. If one hand divides it the other hand simultaneously divides it equally. Nature will not allow her balance to be disturbed. Each polarized hand moves away from the other from zero to four. Each then reverses its polarity from a charging body to a discharging one. Generation also reverses and becomes radiation. Heating bodies cool. Living bodies die. Solids become gases and dissolve. Fast motion slows until it ceases.

THE COSMIC CLOCK

Illustrating the impossibility of disturbing the balance of God's universe, by even the weight of one electron.

If you should cause an explosion in the very center of a perfectly spherical room you would form spherical layers of increasingly dense pressures with maximum density at the surface of the sphere. The center of the sphere would be maximum in vacuity. The explosion would be symmetrically radial. The reaction to that explosion would also be its reverse. The reflections which would return by radar from the spherical walls of that room would collide at its very center. Compression would then be exerted from the outside and density would increase in the direction of the center. Nature does not work that way, however. Nature causes her explosions to take place as though they were confined within the flat walls of a room of four or many walls of such shapes as we see in crystals. If you caused such an explosion in your six sided room the outward expansion would no longer be even. It would not even be spherical because of the four corners, which would have to be filled. The outward explosion could no longer produce straight radial lines, which would reflect back in straight radial lines. Every radial line would have to curve in the direction of its corners, and as they approached those corners their curvature would twist and increase in speed as they approached the corners. In a sphere all radial lines are equal, but in a cube the diagonals are longer than the diameters. This fact accounts for the curvature, the spin and the shaft. It also accounts for the disappearance of all curvature.

If you study figures 19 to 43 for better comprehension of this effect, then realize that Nature does not allow any explosion anywhere without repeating it all over her universe, you will soon become familiar with her simple processes. The eight explosions shown in Fig. 77 are in cube relation to explosive charges placed out in open space in a cubic relation to each other. If they expanded into spheres, such as you see in Fig. 78 there would be a lot of empty space between them which Nature does not allow, so they continue to expand until they fill all space. *Curvature then disappears and so, also, does motion*—as shown in Fig. 79. Every point on every face of those resultant cubes is zero in potential and motion, for they are planes of magnetic Light. They are also mirrors which reverse all direction of motion,

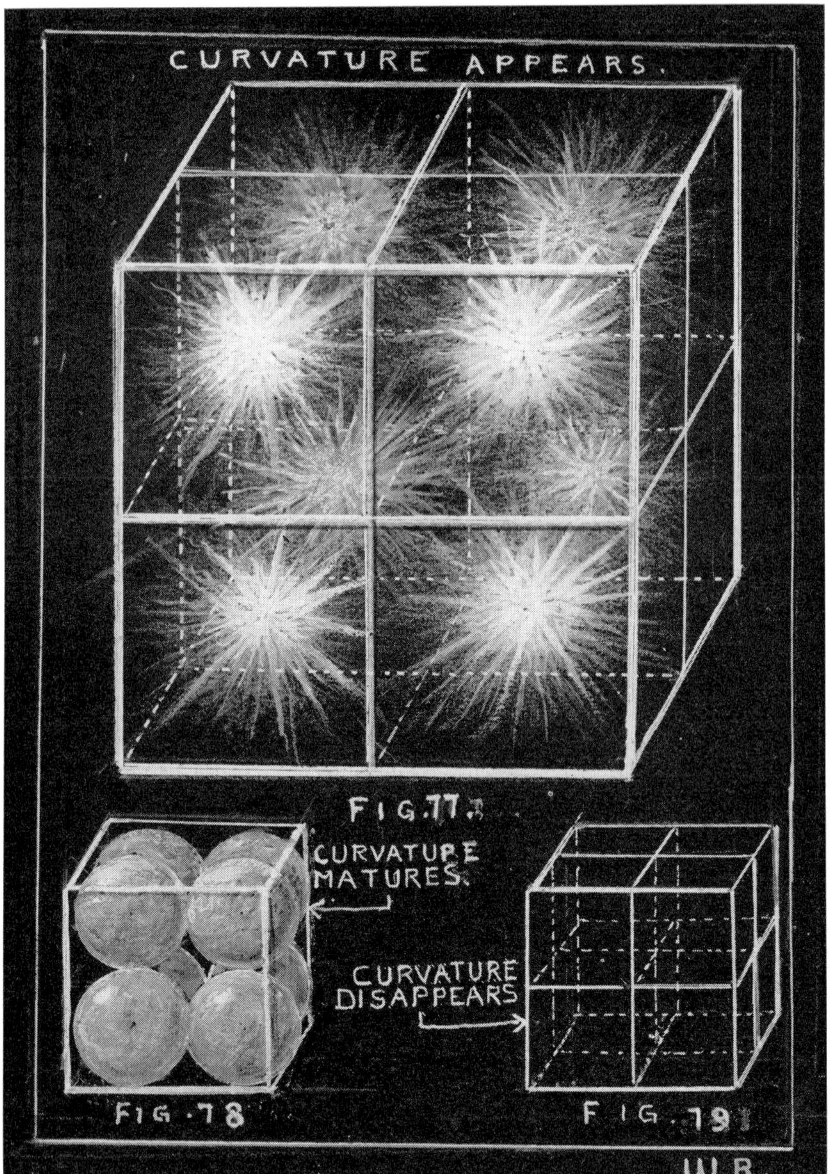

The outward pressures of radiation and centrifugal force lose all spherical form and curvature by disappearance into the six planes of zero curvature, which bound all cubic wave-fields.

which impact against them to extend them into the next wave-field and radar them back to their point of beginning. *When the world of science realizes that radar and polarization are one and the same effect it will greatly advance its mechanics, for it is by this principle that any happening anywhere in the universe happens everywhere. This principle is the whole basis of the telephone, radio, television, etc.*

We have very briefly described this principle which makes polarization and a bilateral universe not only possible but imperative. Likewise, it makes extension imperative. Now we shall apply this principle to very familiar simple effects.

2. Man makes a big gun by making a long tube which he seals at one end and places an explosive there. He even puts a twist in the bore of the gun to accelerate the spin of motion around the still shaft which centers that tube. That motionless explosive occupies one still position in this zero universe until he ignites it. What then happens? A two directional explosion takes place. *The recoil is equal to the discharge,* so he might just as well have left both ends of the gun open as far as the effect of polarization is concerned. The recoil was the opposite direction of polarity, as that explosion began its division of stillness throughout the entire universe, at the speed of 186,400 miles per second.

Let us analyze what has happened. The center of the explosion is an unchangeable point of stillness in the Magnetic Light of Creation, but we will call it gravity because we are seemingly dividing it into a shaft. Because of the tube the explosion cannot expand symmetrically and radially from the point in space which it occupied while still, for it is not enclosed in a sphere. It is enclosed in a shaft. Its spinnings encircle the walls of the tube at ninety degrees to the still shaft, which is developed because the explosion can no longer be centered as one point in space, and, therefore, have but one center of gravity instead of many. It becomes a series of points in space which form a shaft. If you can now comprehend that if the explosion within that tube is obliged to change position and move its centering point of stillness, the projectile which moves is, likewise, a sequence of points of stillness? You cannot see a projectile being ejected from the other end, but its equivalent is in that recoil.

Can you not see that this same thing applies to an airplane? The plane moves for the same reason. An explosive is placed at one end of a tube, which is made up of an engine and a propeller. The engine is sealed at one end because there is no opposing propeller there. If equal propellers occupied both ends of the engine the plane would not move. Spinning rings then move around a center, instead of an extended shaft of gravity, however, caused by the motion of the propellers. The plane itself would then occupy a position of stillness. *Can you not see that it would always occupy a position of stillness, no matter where it seemingly moved?* And can you not see that the spinning rings are the *cause* of that seeming motion? And can you not further see that the direction of motion is not in the projectile, or the plane, or of gravity, *but solely in the divided pressures which cause that spinning motion?*

3. Let us now take the plane and the projectile out of the tubes, which they centered. We refer you to the two cyclone diagrams again—figures 51 and 52. There is nothing there, whatsoever, which is materially objective at their center. There is only the stillness which we call the eye. There is but an explosion of hot air there, which causes expansion from the inside outward where cold air compresses it from the outside to inner density. Nature has made her two-way "gun" that way—a measure of desire-energy is at its center. *The eye is a shaft of gravity.* Gravity is omnipresent. What difference whether there is a plane, or bullet there, or not? *If there is a plane there, its changed position in one direction is voided by an equal change in the opposite polarity direction.*

4. If we go into the machine shop we will see a long shaft with many wheels upon it. They are all turning in the direction of motion yet they are not changing their positions. They perform as much work, or can do as much damage, as the plane or bullet, without changing their gravity position. Which is it, then, which demonstrates power? Is it the direction of motion around stillness, or is it the shifting point of stillness?

5. We now go into the laboratory. The physicist makes a two-way gun, which he calls a coil. He does not seal one end of it for

he cannot. His electric current goes both ways and spins around its extending shaft in equality of potential. In this way he divides gravity and extends it in such a manner that both of each pair of mates, thus divided, are equal.

We shall now apply this principle to the Lee-Yang experiment. In the coil, above described, there is no material body occupying the cathode position as there is when a plane, or bullet, starts to move. In the Lee-Yang experiment a material body is placed there to be polarized. Cobalt is not a balanced, mated pair, like carbon, however. It is but one of a pair, although it is very close to the balanced amplitude point of its wave. When the current is turned on the resultant effect is as though a steel rod was inserted in the shaft of the coil, which is a half inch in diameter at one end and a quarter of an inch at the other end. Every physicist knows that both ends would then be different, for the equator of the rod would not be in the middle. It would be nearer the small end. The rod would not pick up an equal weight at its opposite ends, for the electric potential which does the lifting, would be unequal. The small end would pick up a greater weight, for its rings would be smaller and spin faster.

Another example will help clarify this one. The gun maker builds a two-way gun but does not seal either end. If the tube is of the same dimension from one end to the other, the projectiles ejected from both ends could be the same size, but if he made one end larger than the other the projectiles ejected from one end would be smaller than those ejected from the other end. For this reason we say that carbon, sodium-chloride, or potassium-bromide would give equal projections from each end, for they are mated pairs, whereas sodium or potassium alone would give as unequal results as unequally mated pairs, and sodium-bromide, or potassium-iodide would give as unequal results in such an experiment as their cube crystals are distorted, and for the same reason.

CONCLUSION

As our last words in concluding this explanation of the Lee-Yang experiment, and in trying to explain the universe as a whole, we will draw a crude picture, which should make it more clear, even to the unimaginative.

The universe is made up entirely of electric thought-waves. Thought-waves are Nature's two-way guns, which are loaded with the energy of Mind-desire to give out from its centering Self to all the universe. *That is the explosive which motivates the universe. That is what makes Creation appear and disappear.* If you can imagine these two-way spiral guns, which are shaped like the spirals shown all through this book, and if you can imagine them placed muzzle to muzzle all through the universe, you have one part of the picture clearly in Mind. If you also imagine these two-way spiral guns to be of both atomic and stellar dimensions, you will clarify the picture somewhat. The only thing now to complete the picture is in discharging all of these cosmic guns right into each others muzzles so that the explosive, which is projected, collides as hot bodies, which are frozen into solids as they are ejected from the muzzles of those polarizing guns. That is the way that paired units become mated systems, which complete the purposefulness of the universe by regiving to the Creator all that the Creator had given to them.

Thus it is that The God of Love manifests love to the fathers and mothers of Creation, who, in turn, manifest love by their equal givings and regivings of each to each other as united pairs, and they, in turn, regive of love to their Creator to become One with Him.

Thus it is that matter is created by the process of heating it by fast motion. Thus it is, also, that the cold of space freezes it to imprison it until it is enabled to generate enough heat for its liberation. This fact is not known by man. *He believes that matter is held together from the inside. Because of this belief he does not know that matter is a powder keg which desires to blow up the universe.* It is our belief that man will desist from helping matter to blow up the universe when he is made aware of the Nature of matter as it is, instead of the way he thinks it is. That has been our purpose in writing this book. We reverently launch it out upon the world with the hope and deep prayer that it will save the earth from becoming a barren thing, which it will most surely become, if the deadly blue killer metals are built into ten ton piles sufficiently long to make it forever impossible to turn back to the normalcy required for organic life. *Our last word in closing is to say that a time would be reached within one generation, in which there will be no way to turn back, for the doom of organic life would then be inevitable.*

P.S.

A postscript has become suddenly imperative on the eve of going to press, because of an article published in Scientific American for April, 1957. This article, by Philip Morrison, upholds the validity of the conclusion that the Law of Parity is now overthrown, and that a right and left direction must henceforth be recognized as a part of natural law.

This conclusion is supported mainly by one of many drawings which illustrate the article. We are publishing that drawing to show that the same mistake has been made by this author that Messrs. Lee and Yang made at Columbia University. See Fig. 80.

At the left of the drawing two particles are turning upon their gravity shafts which could be electrons, planets or suns. Around these spinning masses are circles with arrows which show the direction of their turning. Naturally these circles show as an ellipse because they follow equators and are shown in perspective.

At the right of these two spinning masses a mirror has been erected to show that if these two masses were reflected in the mirrors of Nature, such as radar, the spinning would be reversed. Upon this evidence, and similar evidence containing the same fault, a great conclusion of science is now added which upsets the old concept of parity which allowed for no distinction as left and right hand turnings. We have marked this mirror by the letter A.

The great error of this drawing is that pressure mirrors of Nature, such as radar, are not erected upon planes which follow the direction of polarity—or gravity shafts. On the contrary, they are erected in planes of 90 degrees to polarity, or gravity shafts.

We are printing another drawing, Fig. 81, which shows three

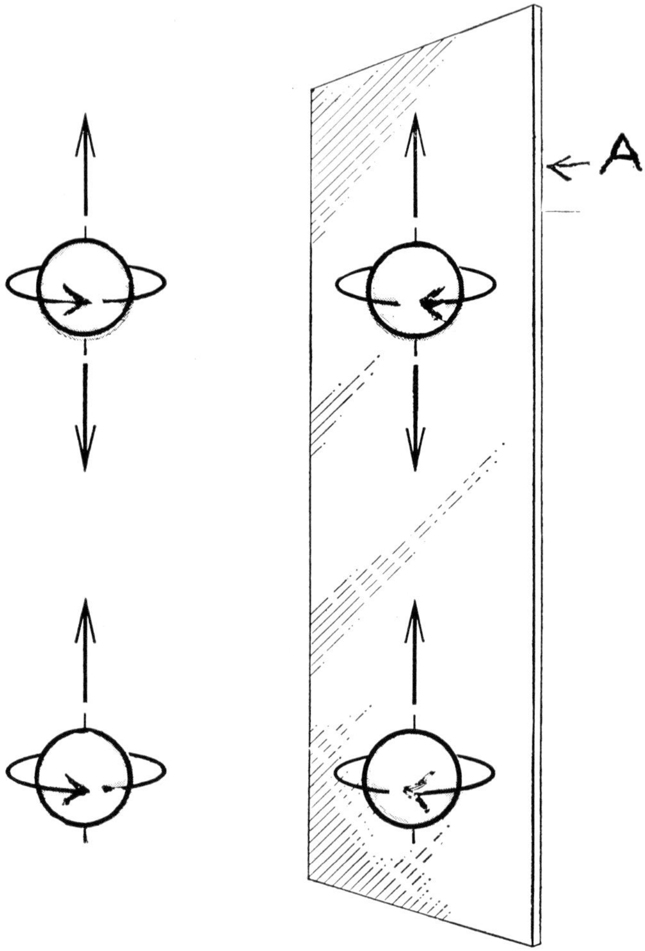

FIG. 80. A mirror placed parallel in plane with the direction of the shaft of gravity, would, naturally, show a reverse direction of spinning. But radar, or echo mirrors of Nature are not placed that way. They are placed at angles of 90 degrees from gravity shaft. See Fig. 81.

mirrors which are placed where Nature places them. These we have marked A—A. If you now note the reflections of the turning masses, which we have indicated upon these mirrors, you can plainly see that the one direction of Nature is upheld.

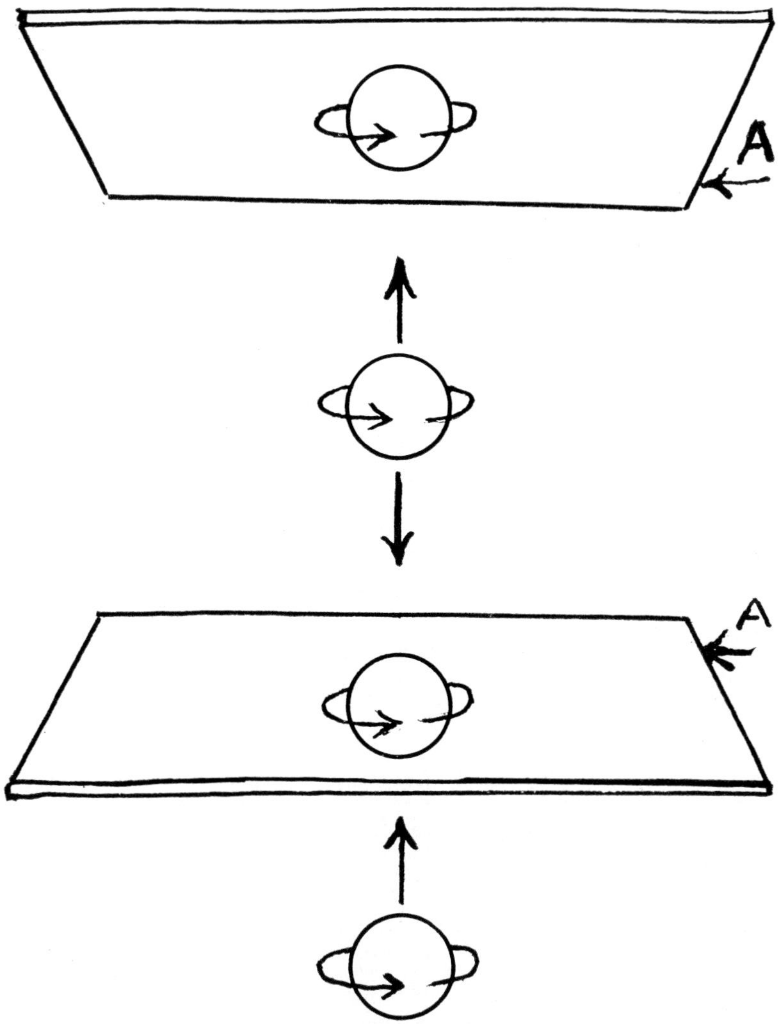

Fig. 81.

FIG. 81. This is the way Nature places her pressure mirrors which retain the same direction of spinning in the mirrors that they spin in the mass which acts upon them. There is no exception to this law throughout all Nature.

To help you visualize this effect more clearly consider how impossible it would be to have a ball rebound into your hand if you threw it against a plane which parallels the direction of your ball. Or consider how impossible it would be for a cliffside to echo your voice if the plane of its face followed the direction of your voice—as it does in the Scientific American drawing—instead of in a plane of 90 degrees from the direction of your voice. Once more consider how impossible it would be for radar instruments to detect the presence of a sheet of steel much bigger than a ship, if that sheet of steel was placed the way the mirror in Fig. 80 is placed, instead of being turned around so that its entire face could echo back to the radar instruments.

Nowhere in Nature do pressure walls exist which are not 90 degrees from the direction of polarity. For this reason, the conclusions arrived at in this article to support the Lee-Yang theory are invalid.

University of Science and Philosophy

A NEW UNIVERSITY IS FORMED IN VIRGINIA, DEVOTED EXCLUSIVELY TO THE SCIENCE OF MAN.

The UNIVERSITY OF SCIENCE AND PHILOSOPHY has just been incorporated in Virginia as the first university ever formed which recognizes that the wealth and power of the world lies in man himself. This new university also recognizes the fact that man is a spiritual creation whom God endowed with inner power to have dominion over all the earth through knowledge of his own inherent power, instead of being but a unit in the world-machine.

It fully recognizes that man is the greatest, and highest, product of God, and that the great need of this chaotic world is to make man know his own value to man through knowing himself, instead of measuring his value as a material producing machine. The marvelous educational facilities of our schools and universities have given man fine and necessary techniques that have enabled him to build a world full of wondrous things in the last hundred years, but he has also built a world of fear and hate which is even now threatening the survival of man himself. The great question of all this is whether we have not gone too dangerously far in the attempt to build luxury for man at the expense of his Soul, by placing material values ahead of spiritual ones? Have we not done what the great poet Edwin Markham feared we were doing when he expressed his fear in these words: "In vain do we build the city if we do not first build the man"? Have we not mistaken encyclopedic information for knowledge, and created great masses of men who REMEMBER and REPEAT instead of men who THINK AND KNOW?

Have we not built our Machine Age by suppressing the immortality of MAN? Alexis Carrel was so fully convinced that we had entirely forgotten MAN in the structure of our civilization that he wrote his grief into MAN THE UNKNOWN. In that book he said: "The Science of Man is the most important, and least known, of all the sciences." The entire purpose of the new University is to awaken the creative genius and cosmic powers which lie dormant in every living being.

It is about time that we give thought to the immortality of man, for world culture has been dangerously declining for the last fifty years because of lack of support and patronage to world-genius in favor producing more, and still more, war and luxury machines.

The UNIVERSITY OF SCIENCE AND PHILOSOPHY is new in name only. It has already existed for ten years in that fine old marble palace and classic sculpture gardens on a mountain top near Waynesboro, Virginia, which Calvin Coolidge once used as a summer White House.

It was founded by Lao Russell under the name of the Walter Russell Foundation in 1948. Through her dedicated leadership and inspired, untiring efforts, she has built a student body of approximately 50,000 students all over the world, who have been studying the Science of Man Course in their homes, which both the Russells have written together, in UNIVERSAL LAW, NATURAL SCIENCE AND LIVING PHILOSOPHY,—also their separately written books which give knowledge of man to man, and also give new knowledge of God's ways and processes in the creation of our universe.

The world knows Lao Russell as a higher philosopher and humanist, but she is also a gifted scientist and sculptor whose colossal marble CHRIST OF THE BLUE RIDGE, which she created with Walter Russell, inspires and illumines the many thousands who visit the mountain top paradise that she has created during her ten years of preparation for the unfolding of this university. Having arrived at this point The Walter Russell Foundation merely "grew up" into its university status.

The world also knows Dr. Walter Russell in the arts, and in

Swannanoa—Headquarters of the University of Science and Philosophy.

Sculpture Gardens of the Works of Walter and Lao Russell.

science as "The Modern Leonardo", "Dean of the Arts", "The most versatile man in America" and "The man who tapped the secrets of the universe", for he has risen to high altitudes as a painter, sculptor, author, composer of music, architect, philosopher, scientist and humanist.

Behind those ten years lies a still greater background of history covering nearly a century in time, and the greatest names in contemporary history, who gave birth to this greatest moral movement of all the centuries, which is now culminating in this new university. Herbert Spencer began it way back in the 1870's. He was deeply concerned about the ability of man to survive and progress in a social system which placed material values first, and trained men to produce material "wealth" at the cost of their Souls. He imparted these worries to his friends the poets, whom he believed to be the wisest and most illumined intellectuals in the world, such men as Ralph Waldo Emerson, Walt Whitman, Edwin Markham, Oliver Wendell Holmes, John Burroughs, Mark Twain, Henry Holt and Richard Watson Gilder, and they in turn impressed Andrew Carnegie with the necessity of building world culture by giving libraries to communities which lacked them. The greatest achievement of this initial group of Founders of the Science of Man movement was the writing of a moral code which was based upon man as the greatest product of God's Creation, and of service of man to man as man's greatest expression of unity, righteousness, good citizenship, good neighborliness and moral character.

That moral code of the poet mystics was threatened with death by inertia twenty years after its birth, through lack of leadership caused by Herbert Spencer's return to England. It was then, in 1895, that Walter Russell gave it its much needed dynamic leadership for many years of conspicuous achievement. He gathered together those poets who still lived, including Edwin Markham, who had deep fears for man, and added such other great contemporary names as Robert Collier, Alexis Carrel, Richard Le Gallienne, John Dewey, Theodore Roosevelt (Rudyard Kipling and Sir Arthur Conan Doyle in England), Dr. Lee de Forest, Dan Beard, Nikola Tesla and many many other intellectuals. During the following fifteen years it was through Walter Russell that this powerfully spiritual group brought into being the Boy Scouts movement in England, through the efforts of Kipling's enthusiasm, and Sir Baden Powell's activities,—then the Boy Scouts of America, through Dan Beard. The Rotary, Kiwanis, Lions and other service clubs soon followed from 1905 on, until today hundreds of millions of men and youths are becoming infiltrated with the principle of brotherhood embodied in that Spencer Code, which all adopted with slight modifications of their own.

Once again the whole movement faced dissolution because of the long German war, but God would not let it die. Thomas Watson so sincerely felt the need of it when he came to New York as the head of International Business Machines, that he made it financially strong in order that it may continue, and spiritually strong by his own magnetic personality. He became its president, and many more great names of the higher intellectuals were added. With that greater strength he turned back the leadership to Walter Russell who again held the Presidency for seven years of continued reforms and cultural achievement of great value to mankind. Noteworthy among these were the sponsoring of the Bronx River Parkway by awarding its engineers with gold medals of honor,—sweat shop elimination—compulsory schooling—advertising censorship (headed by Adolph Ochs, owner of the New York Times)—Child Welfare (headed by Sophie Irene Loeb and financed by August Hecksher)—better banking ethics (headed by Francis Sisson of the Guaranty Trust Co.)—Better Business Bureau—the writing of MAN THE UNKNOWN by Alexis Carrel and THE DECLINE OF THE WEST by Oswold Spengler, as warnings to our civilization, which has never heeded them—and the awarding of many medals for intellectual service to man.

For twelve years Walter Russell and Thomas Watson worked together in spreading the Spencer Code of ethics and the Watsonian philosophy which grew out of it into the entire business world and raised its standards from the "let the purchaser beware" principle of 1925 to "the mutual service first" principle, in the short period of ten years. No better example of the business teachings in the IBM organization could be cited than the following: *"This is a man age. To make a business grow begin growing men. We are all one brotherhood. We have but one thought, one creed, mutual helpfulness to each other.*

Once again, however, the long continued second world war so weakened the power of this organization—then known as THE SOCIETY OF ARTS AND SCIENCES—it was forced to cease its activities. Walter Russell, having divided his time between his profession and his humanist movement, to which he had also given most of his personal earnings, it became necessary to give his full time to the arts.

Once again God would not let die. In 1946 Lao Russell came into Walter Russell's life. Inspired by the necessity for the survival and perpetuation of this epochal movement, she conceived the idea of a university whose sole teachings would be The Science of Man, and for the past ten years has given of her genius, and the great stimulus of her indomitable spirit and love for humanity, the movement needed for its survival.

This briefly is the story of the University of Science and Philosophy—the first university ever to be founded whose sole purpose is the awakening of that divine spark of man's spiritual nature, that alone can give birth to a united and enduring civilization, with its desired pinnacle of peace, happiness and prosperity, which all men eternally seek.